W9-BCM-383

SOCIAL RESEARCH

USING MICROCASE

Michael Corbett
BALL STATE UNIVERSITY

Lynne Roberts
MICROCASE CORPORATION

SECOND EDITION

Editor	David Smetters
Editorial Assistant	Julie Aguilar
Production Manager	Jodi Gleason
Software, Lead Developer	David H. Simmons
Software, Programmer	Dana Schlatter
Data Archivists	Meredith Reitman
	Chris Bader
Cover/Interior Design	Michael Brugman Design
Copy Editor	Margaret Moore

© 1998 by MicroCase Corporation, All rights reserved.

No part of this book may be reproduced, stored in a retrieval system, or transcribed, in any form or by any means, electronic, mechanical, photocopying, recording, or otherwise, without the prior permission of the publisher, MicroCase Corporation, 14110 NE 21st Street, Bellevue, WA 98007, (425) 641-2506.

Windows 95 is a registered trademark of Microsoft Corporation.

MicroCase and **ExplorIt** are registered trademarks of MicroCase Corporation.

Printed in the United States of America

1 2 3 4 5 6 7 8 9 10–01 00 99 98

Contents

About the Authors

Michael Corbett received his Ph.D. from the University of Iowa. He is now professor of political science at Ball State University, where he teaches courses covering an introduction to political science, research methods in political science, and public opinion. He is author of numerous scholarly articles and of four books: *Political Tolerance in America: Freedom and Equality in Public Attitudes* (New York: Longman, 1982); *American Public Opinion: Trends, Processes, and Patterns* (New York: Longman, 1991); *Research Methods in Political Science: An Introduction Using MicroCase* (Bellevue: MicroCase, 1996); and *Politics and Religion in the United States* (Garland Press, 1998), which was co-authored with his wife, Julia.

Lynne Roberts received her Ph.D. from Stanford University where she also was on the staff of the Computing Center. She left Stanford to inaugurate a year-long research methods course in the then-new doctoral program offered by the School of Social Welfare at the University of California, Berkeley. After several years, she joined the faculty of the University of Washington in Seattle where she served as Associate Professor of Sociology. In 1984, she left academic life to pursue her interests in educational software. Currently, she is president of MicroCase Corporation. During her tenure at Washington, Roberts taught statistics at both the undergraduate and graduate levels. Trained as an experimental social psychologist, she designed and conducted many experiments, the results of which she reported in a book and a number of scholarly articles. She also has published articles on methodological subjects including scaling and experimental design.

Getting Started

INTRODUCTION

This workbook is about *doing* social research. There is nothing make-believe about what you will be doing. You'll use the same resources and techniques used by professional social science researchers.

Since the exercises differ in length and difficulty, your instructor may assign only selected questions, or you may be given alternative questions. Be sure to carefully check any information from your instructor so that you will do the appropriate problems. Some of the problems will ask you to print certain results and attach them to your worksheet. If you do not have a printer or you have been instructed not to use a printer, simply complete the written exercises.

Many of the exercises will have a variety of possible answers. This reflects the nature of research—there is no "correct" way to do research. In most situations, a variety of approaches are equally appropriate.

The exercises are designed so that the introductory material does not require a computer. If your computer time is limited, you may read this material beforehand and save your computer time for completing the worksheets. In any case, you should carefully read the introductory material before starting on the written exercises.

SYSTEM REQUIREMENTS

Two versions of Student MicroCase have been provided with this book: a Windows 95 version and a DOS version. The Windows 95 version can be used only on computers running Windows 95 (or higher). The DOS version of the software will run on most any DOS or Windows compatible computer, including those using Windows 3.1 and Windows 95. Here is more detailed information about the minimum computer requirements for each version of the student software.

- **Student MicroCase for Windows 95**—This version will run on almost any computer having Windows 95 (or higher).[1] For installation purposes, a CD-ROM drive and a 3.5" floppy drive are also required.

- **Student MicroCase for DOS (or Windows 3.1)**—This version requires an IBM or compatible computer with 286 or better processor, 640K RAM, DOS 3.1 or higher (or Windows 3.1 or higher), VGA-level graphics, and a mouse. A CD-ROM is *not* required.

Before installing any software to a hard drive, check with your instructor to see if a network version of Student MicroCase has already been installed. If a network version has already been installed, skip to the section "Starting Student MicroCase."

    If the operating system on your computer is DOS or Windows 3.1, then you must use *Student MicroCase for DOS*. If this is your case, skip to the section "Starting Student MicroCase."

    Even if your computer has Windows 95 (or higher), there are some situations in which you should still use *Student MicroCase for DOS*:

- your teacher has instructed you to use the DOS version (yes, the DOS version can be used on Windows 95 computers)

- you are not allowed to install software to the hard drive of the computer (such as in a lab setting)

- your computer does not have a CD-ROM drive

- your computer does not have a CD-ROM drive and a floppy drive that can be used at the same time (as with some notebook computers).

If any of these conditions apply, you should skip to the section "Starting Student MicroCase."

NETWORK VERSIONS OF STUDENT MICROCASE

Network versions are available for both the Windows 95 and DOS versions of Student MicroCase. These special versions of the software are available at no charge to instructors who adopt this book for their course (instructors should contact MicroCase Corporation for additional information). It's worth noting that *Student MicroCase for DOS* can be run directly from the floppy diskette on virtually any computer network—regardless of whether a network version of Student MicroCase has been installed.

---

[1] *Student MicroCase for Windows 95* requires 8 megabytes of RAM, 10 megabytes of free hard disk space (network installations require around 1 megabyte of temporary storage on hard drives of user terminals), VGA-level graphics, and a mouse.

INSTALLING STUDENT MICROCASE FOR WINDOWS 95

If you will be using *Student MicroCase for DOS* (see above discussion), you do not need to read this section. Skip to "Starting Student MicroCase."

In order to install *Student MicroCase for Windows 95*, you will need the floppy diskette and CD-ROM that is packaged inside the back cover of this book. Then follow these steps:

1. Start your computer and wait until the Windows 95 desktop is showing on your computer.

2. Insert the floppy diskette into the A drive (or B drive) of your computer.

3. Insert the CD-ROM disc into the CD-ROM drive.

4. Click on [Start] from the Windows 95 desktop, click [Run], type **D:\SETUP** (if your CD-ROM drive is not the D drive, substitute the letter D with the proper drive letter), and click [OK].

5. During the installation, you will be presented with several screens (described below). In some cases you will be required to make a selection or entry and then click [Next] to continue.

The first screen that appears is the **Welcome** screen. This provides some introductory information and suggests that you shut down any other programs that may be running. Click [Next] to continue.

You are next presented with a **Software License Agreement**. Read this screen and click [Yes] if you agree with the software license.

If this is the first time you are installing Student MicroCase, an **Install License** screen appears. (If this software has been previously installed or used, then it already contains the licensing information. A screen simply confirming your name will appear instead.[2]) Here you are asked to type in your name. It is important to type your name correctly, since it cannot be changed after this point. Your name will appear on all printouts, so make sure you spell it completely and correctly! Then click [Next] to continue.

The next screen has you **Choose the Destination** for the program files. You are strongly advised to use the destination directory that is shown on the screen. Click [Next] to continue.

---

[2] If an installation of Student MicroCase—including those versions distributed with other MicroCase textbooks—is already on your hard drive, you will get a warning message indicating that a copy of the program already exists on your computer. If your intention is to *replace* the previously installed version of Student MicroCase, it is recommended that you exit the install program, remove the earlier installation using "Add/Remove Programs" from the Windows 95 Control Panel, and then reinstall Student MicroCase. Otherwise a second installation of Student MicroCase will be placed on your hard drive—that is, the original installation will not be replaced. If you previously installed a version of Student MicroCase for another product on your hard drive, go ahead and continue with this installation—again, the previous installation will not be overwritten.

The **Install Checkbox** screen requires you to make a choice about whether or not to copy the data files (currently located on the floppy diskette) to your hard drive. Carefully read the choices on the screen before making your selection.

When the **Setup Complete** window appears click [Finish]. You will find it easier to start Student MicroCase if you place a "shortcut" icon on your Windows desktop. A folder named "MicroCase" should now be showing on the horizontal task bar at the bottom of your Windows desktop. Click on this button and a window will appear with a "shortcut" icon for Student MicroCase.[3] Place your mouse pointer over this icon, then press down *and* hold the left mouse button as you drag the icon outside the window to an open space on your Windows desktop. Once the Student MicroCase icon has been moved to your desktop, you can close the window that previously contained the icon by clicking the little "x" button that appears in the top right corner of the window. From this point on, you will be able to double-click the Student MicroCase shortcut icon to start the software.

STARTING STUDENT MICROCASE

The first section below describes how to start *Student MicroCase for Windows 95,* while the second section describes how to start *Student MicroCase for DOS.* Read the section that is appropriate for you.

STARTING STUDENT MICROCASE FOR WINDOWS 95

Student MicroCase for Windows 95 must first be installed to a hard drive (or a computer network) before you can start it. If the program has not been installed, review the software installation section above.

If the data files were not copied to the hard drive of the computer during the installation of *Student MicroCase for Windows 95,* it will be necessary for you to insert your 3.5" data file diskette into the A or B drive of your computer. (If you are starting Student MicroCase from a network, you *must* insert your floppy diskette before continuing.) Don't worry, you will be prompted to insert your floppy diskette if you forget.

If the software was installed properly, there should be a "shortcut" icon on your Windows desktop that looks something like this:

---

[3] If you have another installation of Student MicroCase on your hard drive, make sure to rename the label for its "shortcut" icon on your Windows desktop before following the instructions in the next sentence (you can name it anything *except* "Student MicroCase"). To rename a shortcut, click once on the shortcut label, and then click it again. This causes the text for the shortcut label to be highlighted, which can then be modified.

To start *Student MicroCase for Windows 95,* position your mouse pointer over the shortcut icon and give it a double-click (that is, click it twice in rapid succession). If you did not move the shortcut icon onto your desktop during the install process, you can alternatively follow these directions to start the software.

Click [Start] from the Windows 95 desktop.

Click [Programs].

Click MicroCase.

Click Student MicroCase.

After a few seconds, Student MicroCase should appear on your screen. Skip down to the "Main Menu of Student MicroCase" section below to continue your introduction to the software.

STARTING STUDENT MICROCASE FOR DOS

This section explains how to start *Student MicroCase for DOS.* You can run *Student MicroCase for DOS* directly from the floppy diskette on almost any DOS or Windows computer (including computers using Windows 3.1 or Windows 95).

The instructions for starting *Student MicroCase for DOS* differ depending on the operating system you are using. In all cases, you will first need to place the 3.5" floppy diskette in the A or B drive. Go ahead and do that now. Then follow the appropriate instructions below to start *Student MicroCase for DOS* on your computer.

MS-DOS
Type **A:MC** (or **B:MC**) and press <Enter>.

Windows 3.1
From the Program Manager click on [File].
Click on [Run].
Type **A:MC** (or **B:MC**) and click [OK].

Windows 95
Click on [Start].
Click on [Run].
Type **A:MC** (or **B:MC**) and click [OK].

The first time you start *Student MicroCase for DOS,* you will be asked to enter your name. It is important to type your name correctly, since it will appear on all printouts. Type your name and click [OK] or press <Enter>. If your name is correct, simply click [OK] or press <Enter> in response to the next prompt. (If you wish to correct a mistake, click on [Cancel] to make a correction.)

You will notice from the opening copyright screen that the full name for the DOS version of this software is "MicroCase ExplorIt." But just to keep things simple, it will be referred to as "Student MicroCase" throughout this book so that it matches the name for its Windows counterpart. To continue to the main menu of the program, press <Enter> or click the left mouse button.

Note: If you are using Windows 3.0 or 3.1 and the mouse fails to appear or it does not work properly, refer to Appendix A.

MAIN MENU OF STUDENT MICROCASE

Student MicroCase is extremely easy to use. All you do is "point and click" your way through the program. That is, use your mouse arrow to point at the selection you want, and then click the left button on the mouse. The main menu is the starting point for everything you will do in Student MicroCase. Let's take a look at how it works.

Student MicroCase for Windows 95—The Windows 95 version of Student MicroCase works from two different menus: the FILE & DATA MENU and the STATISTICS MENU. You can toggle back and forth between these two menus by clicking the menu names shown on the left side of the screen. All options for each menu are not always available. You will know which options are available at any given time by looking at the color of the options. For example, when you first start the software, only the OPEN FILE and NEW FILE options are immediately available. As you can see, the colors for these two options are colored more brightly than the others shown on the screen. Also, when you move your mouse pointer over these two options, they become highlighted.

Student MicroCase for DOS—The DOS version of Student MicroCase works from a single main menu. When you are at the main menu, only those tasks having a bright yellow background are available. As you can see, no tasks are available until you select a data file with which to work.

SOFTWARE GUIDES

Throughout this workbook, there are "Software Guides" that provide you with the basic information needed to carry out each task. Here is an example:

➤ *Data File:* **USA**
   ➤ *Task:* **Mapping**
➤ *Variable 1:* **109) MURDER**
   ➤ *View:* **Map**

Each line of this Software Guide is actually an instruction. Let's follow the simple steps to carry out this task.

STEP 1: SELECT A DATA FILE

Before you can do almost anything in Student MicroCase, you need to open a data file.

Student MicroCase for Windows 95—To open a data file, click the OPEN FILE task from the FILE & DATA MENU. A list of data files will appear in a window (e.g., GSS96, GSS93, USA, etc.). If you click on a file *once*, a description of the highlighted file is shown in the window next to this list. In the Software Guide shown above, the ➤ symbol to the left of the *Data File* step indicates that you should open the USA data file. To do so, click on USA and then click the [Open] button (or, just double-click on USA). The next window that appears (labeled *File Settings*) provides additional information about the data file, including a file description, the number of cases in the file, and the number of variables, among other things. To continue, click the [OK] button. You are now returned to the main menu of Student MicroCase. (You won't need to repeat this step until you want to open a different data file.) Notice that you can always see which data file is currently open by looking at the file name shown on the top line of the screen.

Student MicroCase for DOS—In the DOS version of Student MicroCase, the available data files are listed in the window at the left of the screen, and the description of the highlighted file is shown in the window beneath this list. To see the description of a file, click on it. To select a file, double-click on its name. The "x" in the box next to the name of the file indicates which file is open. In this example, you should open the USA data file. (You won't need to repeat this step until you want to use a different data file.)

STEP 2: SELECT A TASK

Once you have selected a data file, the next step is to select a program task (if you are using *Student MicroCase for Windows 95*, first switch to the STATISTICS MENU). There are eight analysis tasks offered in this version of Student MicroCase. Not all tasks are available for each data file because some tasks are appropriate only for certain kinds of data. MAPPING, for example, is a task that applies only to ecological data, and thus cannot be used with survey data files.

In the Software Guide we're following, the ➤ symbol on the second line indicates that the MAPPING task should be selected, so click the MAPPING option with your left mouse button.

STEP 3: SELECT A VARIABLE

After a task is selected, you will be shown a list of the variables in the open data file. Notice that the first variable is highlighted and a description of that variable is shown in the Variable Description window at the lower right. You can move this highlight through the list of variables by using the up and down cursor keys (as well as the <Page Up> and <Page Down> keys). You can also click once on a variable name to move the highlight and update the variable description. Go ahead—move the highlight to a few other variables and read their descriptions.

If the variable you want to select is not showing in the variable window, use the scroll bars located on the right side of the variable list window to move through the list. See the following figure:

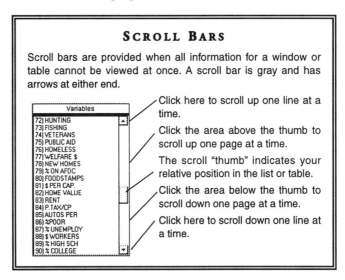

By the way, there is an appendix section at the back of this workbook that contains a list of the variable names for key data files provided in this package.

Each task requires you to select one or more variables, and the Software Guides indicate which variables should be selected. The Software Guide example here indicates that you should select 109) MURDER as Variable 1. On the screen, there is a box labeled Variable 1. Inside this box there is a vertical cursor that indicates that this box is currently an active option. When you select a variable, it will be placed in this box. Before selecting a variable, be sure that the cursor is in the appropriate box. If it is not, place the cursor inside the appropriate box by clicking the box with your mouse. This is important because in some tasks the Software Guide will require more than one variable to be selected, and you want to be sure that you put each selected variable in the right place.

To select a variable, use any one of the methods shown below. (Note: If the name of a previously selected variable is in the box, use the <Delete> or <Backspace> key to remove it—or click the [Clear All] button.)

- Type in the **number** of the variable and press <Enter>.

- Type in the **name** of the variable and press <Enter>. Or, you can type just enough of the name to distinguish it from other variables in the data—MUR would be sufficient for this example.

- Double-click on the desired variable in the variable list window. This selection will then appear in the variable selection box. (If the name of a previously selected variable is in the box, the newly selected variable will replace it.)

- In *Student MicroCase for Windows 95*, you have a fourth way to select a variable. First, highlight the desired variable in the variable list, then click the arrow that appears to the left of the variable selection box. The variable you selected will now appear in the box. (If the name of a previously selected variable is in the box, the newly selected variable will replace it.)

Once you have selected your variable (or variables), click the [OK] button to continue to the final results screen.

STEP 4: SELECT A VIEW

The next screen that appears shows the final results of your analysis. In most cases, the screen that first appears matches the "view" indicated in the Software Guide. In this example, you are instructed to look at the Map view—that's what is currently showing on the screen. In some instances, however, you may need to make an additional selection to produce the desired screen.

MURDER -- 1992: HOMICIDES PER 100,000 POPULATION (UCR)

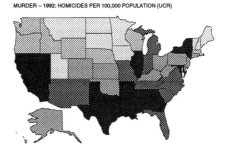

(OPTIONAL) STEP 5: SELECT AN ADDITIONAL DISPLAY

Some Software Guides will indicate that an additional "Display" should be selected. In that case, simply click on the option indicated for that additional display. For example, this Software Guide might have included an additional line that required you to select the Legend display.

STEP 6: CONTINUING TO THE NEXT SOFTWARE GUIDE

Some instructions in the Software Guide may be the same for at least two examples in a row. For instance, after you display the map for murder in the example above, the following Software Guide might be given:

> *Data File:* **USA**
>   *Task:* **Mapping**
> ➤ *Variable 1:* **106) CRIME RATE**
>   ➤ *View:* **Map**

Notice that the first two lines in the Software Guide do not have the ➤ symbol located in front of the items. That's because you already have the data file USA open and you have already selected the MAPPING task. With the results of your first analysis showing on the screen, there is no need to return to the main menu to complete this next analysis. Instead, all you need to do is select CRIME RATE as your new variable. If you are using *Student MicroCase for Windows 95*, click the [[⬒]] button located in the top left corner of your screen (if you are using *Student MicroCase for DOS*, click the [Exit] button once). The variable selection screen for the mapping task appears again. Replace the variable with 106) CRIME RATE and click [OK].

To repeat: You only have to do those items in the Software Guide that have the ➤ symbol in front of them. If you start from the top of the Software Guide, you're simply wasting your time.

If the Software Guide instructs you to select an entirely new task or data file, you will need to return to the main menu. To return to the main menu using *Student MicroCase for Windows 95*, simply click the [Menu] button located at the top left corner of the screen. To return to the main menu using *Student MicroCase for DOS*, click the [Exit] button until the main menu appears. At this point, select the new data file and/or task that is indicated in the Software Guide. (Remember, if you are using *Student MicroCase for Windows 95*, you will also need to switch to the **FILE & DATA MENU** before you can select a new data file.)

That's all there is to the basic operation of Student MicroCase. Just follow the instructions given in the Software Guide and point and click your way through the program.

EXITING FROM STUDENT MICROCASE

If you are continuing to the next section of this workbook, it is *not* necessary to exit from Student MicroCase quite yet. But when you are finished using the program, it is very important that you properly exit the software—do not just walk away from the computer or remove your diskette. To exit Student MicroCase, return to the main menu and select the [Exit] button that appears on the screen.

Important: If you inserted your floppy diskette prior to starting Student MicroCase, remember to remove it before leaving the computer.

Introductory Exercise: Exploring Data Files

OVERVIEW

In this brief, introductory exercise, you will discover how easy it is to use MicroCase to explore two of the data files that you will be using throughout this workbook. Make sure you have already gone through the *Getting Started* section that is provided in the previous section of the workbook. It contains important information that is essential for doing this and future exercises.

In this exercise you will learn to:

- use the USA data file to map variables;
- use the GSS96 data file to create pie charts;
- use the search feature to find any variables that contain a particular word or phrase;
- print the results of your analysis.

There is no chapter in the textbook that corresponds to this introductory exercise. Thus, you may complete this exercise before you begin reading the textbook.

MICROCASE DATA FILES: A QUICK TOUR

In the *Getting Started* section, you learned how to follow the "Software Guides" in order to complete the required analysis task. Let's take this a step further and look at several additional features of Student MicroCase and explore some of the data files that you will be using. Other features in Student MicroCase will be explained later as you need them.

MAPPING

Follow the instructions in the MicroCase Guide below.

> ➤ *Data File:* **USA**
> ➤ *Task:* **Mapping**
> ➤ *Variable 1:* **72) HUNTING**
> ➤ *View:* **Map**

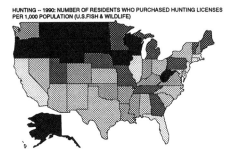

HUNTING -- 1990: NUMBER OF RESIDENTS WHO PURCHASED HUNTING LICENSES PER 1,000 POPULATION (U.S.FISH & WILDLIFE)

Below the Software Guide, you will sometimes be provided additional information or tips about how to create the requested analysis. This information will appear in bold, like it is here. For this example, remember that the first line indicates that you should open the USA data file, the second line tells you to select the MAPPING task, the third line indicates that you should select 72) HUNTING as Variable 1 and the final line indicates that your final view should be the map.

Note that the full description of the 72) HUNTING variable is as follows: NUMBER OF RESIDENTS WHO PURCHASED HUNTING LICENSES PER 1,000 POPULATION. A variable is anything that varies among the objects being examined. Since we are dealing with states in the USA data file, a variable here would be something that varies across states. For example, the crime rate is one such variable. Church membership rate, voter participation rate, population, and geographic area are other variables included in this data file. In this course, we'll spend quite a bit of time talking about variables, how to create them, and how to use them. Here we will see how this variable 72) HUNTING varies across the states.

At this point you have a map of the United States on the screen similar to the one shown above. The map shows the hunting license rate for each state. The states appear in five different colors, from very light to very dark. The darker the state, the higher the hunting license rate. As you can see, there are patterns here. In general, the hunting license rates are higher in the northwest and the upper midwest.

Let's explore some of the other options on the screen by using the following Software Guide.

Data File: **USA**
Task: **Mapping**
Variable 1: **72) HUNTING**
View: **Map**
➤ Display: **Find: Minnesota**

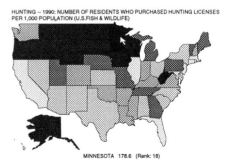

HUNTING -- 1990: NUMBER OF RESIDENTS WHO PURCHASED HUNTING LICENSES PER 1,000 POPULATION (U.S.FISH & WILDLIFE)

MINNESOTA 178.6 (Rank: 18)

Notice that only the last step has a ➤ symbol located in front of it. That means that the first three steps are not required if you are continuing from the previous example. With the hunting license map showing on the screen, you need only to click the [Find case] option. A list of the states will appear. Select Minnesota from the list (*Student MicroCase for DOS* requires a double-click). Then click [OK]. The map for Minnesota has now been highlighted and information about it is presented on the screen.

We see that Minnesota has 178.6 hunting licenses per 1,000 population and is ranked 18th in the nation in terms of this variable.

You can also see the information for a particular state simply by clicking on that state on the map. Try this with several states. Locate some states that are ranked very high on this hunting license rate by clicking on states that have very dark colors. Also locate some states that are ranked very low by clicking on states that have very light colors.

To see all 50 states ranked from highest to lowest, we need to use MicroCase's [List: Rank] option.

Data File: **USA**
Task: **Mapping**
Variable 1: **72) HUNTING**
➤ View: **List: Rank**

RANK	CASE NAME	VALUE
1	Idaho	586.4
2	North Dakota	575.3
3	Montana	569.8
4	South Dakota	538.2
5	Wyoming	470.9
6	Oregon	351.7
7	West Virginia	339.4
8	Wisconsin	302.2
9	Iowa	263.4
10	Alaska	254.2

Again, notice that only the final step must be completed if you are continuing from the previous example. In this case, click the [List: Rank] option (in *Student MicroCase for DOS*, select [List] then make sure the [Rank] option is selected).

You can see from this that Idaho is highest with a rate of 586.4 hunting licenses per 1,000 population and Hawaii is lowest with a rate of 11.7. Click on the [Map] option to return to the map and then do the final step in this Software Guide.

Data File: **USA**
Task: **Mapping**
Variable 1: **72) HUNTING**
➤ View: **Map**
➤ Display: **Legend**

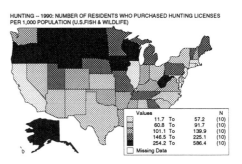

The legend shows you what the colors on the map represent. You can see, for example, that states with hunting license rates between 254.2 and 586.4 are in the highest group, while states with rates from 11.7 to 57.2 are in the lowest group. To remove the legend from the screen simply click the [Legend] option again.

Student MicroCase offers yet a second way to view the map, which you can see by completing the final step shown in this Software Guide.

Data File: **USA**
Task: **Mapping**
Variable 1: **72) HUNTING**
View: **Map**
➤ Display: **Spot**

Each state's hunting license rate is represented by a spot. The size of the spot is determined by the relative magnitude of the value of each state for the hunting license rate variable. Thus, states that have high hunting license rates, such as Idaho and North Dakota, have large, dark-colored spots. State that have low hunting license rates, such as Hawaii and Rhode Island, have small, light-colored spots.

Let's look at the map of another variable. Follow the instructions in the Software Guide below, but as you do, remember that you can select a variable in any one of several ways: by typing in the name of the variable, by typing the number of the variable, or by selecting the variable from the variable list. Try this using each of the three methods.

Data File: **USA**
Task: **Mapping**
➤ *Variable 1:* **70) PICKUPS**
➤ *View:* **Map**

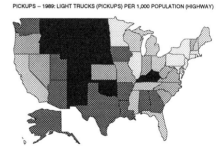

PICKUPS -- 1989: LIGHT TRUCKS (PICKUPS) PER 1,000 POPULATION (HIGHWAY)

If you do not remember how to return to the variable selection screen, you should probably review the *Getting Started* section again. Remember, it is not necessary to return to
the main menu since you do not need to select a new data file or analysis task.

This is a map of the number of pickups per 1,000 population. Let's rank the
states to see which states have the highest rate of pickup trucks.

Data File: **USA**
Task: **Mapping**
Variable 1: **70) PICKUPS**
➤ *View:* **List: Rank**

RANK	CASE NAME	VALUE
1	Idaho	363.6
2	South Dakota	347.2
3	Wyoming	333.3
4	Montana	310.1
5	North Dakota	292.0
6	New Mexico	288.9
7	Oklahoma	242.7
8	Colorado	230.2
9	Kentucky	221.9
10	Nebraska	221.5

We see that Idaho ranks the highest with a rate of 363.6 pickups per 1,000
population. This isn't much of a surprise. In fact, the map of pickup trucks looks
very similar to the map of hunting licenses.

Student MicroCase allows you to print all results that you obtain. If you are
using *Student MicroCase for Windows 95*, you will see a printer icon on the tool bar.
If you are using *Student MicroCase for DOS*, you will see a [Print] button instead. If
you click the print button, a window will appear that gives you the option to
print out the result on the screen. (You can also modify the printer settings, if necessary). Fancy graphics, such as maps, take longer to print than, say, a table or a
page of text. Throughout this workbook, you will be required to print out results
and turn them in with your worksheet assignments. Incidentally, when you print
out a result from Student MicroCase, your name and the current date will appear
at the top of each page. This allows you to easily locate your printout if you are
using a shared printer.

Let's look at another map.

Data File: **USA**
Task: **Mapping**
➤ *Variable 1:* **37) WARM WINTR**
➤ *View:* **Map**

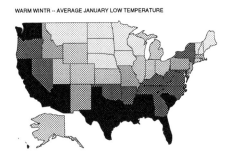

This is a map of the average low temperature in January. This map is almost exactly the opposite of the map of hunting license rates. Not too surprising—the best opportunities for hunting tend to be in the colder states. If you select the [List: Rank] option, you will see that Hawaii is the warmest state with an average January low of 65, while North Dakota is the coldest with an average of –3 degrees.

Now let's see how easy it is to compare two maps in MicroCase.

Data File: **USA**
Task: **Mapping**
Variable 1: **37) WARM WINTR**
Variable 2: **72) HUNTING**
➤ *Views:* **Map**
➤ *Display:* **Spot**

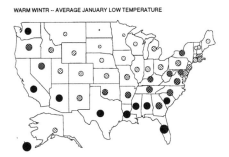

Note that you do not have to modify Variable 1 (WARM WINTR) if you are continuing from the previous example. But you do need to select HUNTING as Variable 2. To do so, click once in the Variable 2 box to make it active, then select the HUNTING variable as you normally would. When the map appears, make sure to select the spot display as indicated.

You can now compare the map of winter temperatures with the map of the hunting license rate.

Let's look at one last map before we leave this data file.

Data File: **USA**
Task: **Mapping**
➤ Variable 1: **101) %NO RELIG**
➤ View: **Map**

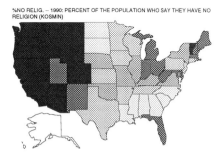
%NO RELIG. -- 1990: PERCENT OF THE POPULATION WHO SAY THEY HAVE NO RELIGION (KOSMIN)

This map shows the percent of the population claiming no religion. Notice that on this map Alaska and Hawaii are not assigned one of the colors. Let's list the values for each state so we can see what is going on with these two states.

Data File: **USA**
Task: **Mapping**
Variable 1: **101) %NO RELIG**
➤ View: **List: Rank**

RANK	CASE NAME	VALUE
1	Oregon	17.2
2	Washington	14.0
3	Wyoming	13.5
4	Nevada	13.2
5	California	13.0
6	Arizona	12.2
7	Idaho	11.9
8	Vermont	11.4
8	Colorado	11.4
10	Montana	10.2

If you scroll to the bottom of the list, you will find Alaska and Hawaii. No data were available for these two states, and hence, they have what is called "missing data." As you will see, it is not unusual to have missing data for a variable, and this can sometimes cause problems for statistical analysis.

UNIVARIATE STATISTICS

Next we will explore a different data file—one based on responses from individuals in a survey—and we will look at some different MicroCase features. As shown in the Software Guide below, open the GSS96 data file and select the UNIVARIATE task.

➤ *Data File:* **GSS96**
     ➤ *Task:* **Univariate**

These data are from the 1996 General Social Survey conducted by the National Opinion Research Center. There were 2,904 respondents (or cases), and this data file contains over 100 variables from that survey. Let's look at some of the variables in this data file. The first variable is the respondent's sex. When the categories of a variable are labeled, these labels are shown in the variable description window. Thus, there are two categories of the variable sex: male and female. For another example, go down to the variable 9) CRIME $ and highlight it. Note that there are three categories for this spending attitude variable: Too little, the right amount, and too much.

Scroll through the list of variables and highlight any that look of interest to you. Examine each variable description to obtain further information about the variable. Some of these variables relate to characteristics of the respondents such as age, sex, and race. These are called *demographic* characteristics. Other variables are about attitudes and behaviors, such as attitudes toward abortion and voting behavior.

Also note that the variables in this data file are based on individual people rather than on geographic areas such as states. Thus, we cannot use the MAPPING task for individuals because you cannot "map people." So we will use the UNIVARIATE task to begin our look at these data. Let's select the variable SEX, as shown in this Software Guide.

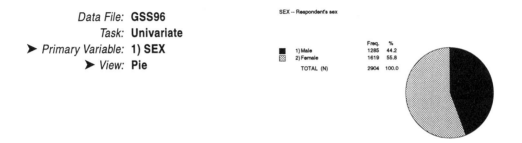

Data File: **GSS96**
Task: **Univariate**
➤ Primary Variable: **1) SEX**
     ➤ View: **Pie**

SEX -- Respondent's sex

		Freq.	%
■	1) Male	1285	44.2
▨	2) Female	1619	55.8
	TOTAL (N)	2904	100.0

A pie chart appears and a legend for the pie colors is shown at the left. We can see that the slice of the pie representing females is larger than the slice representing males. (In chapter 4 of the text, we will see that the sample used in this survey has a slight sex bias: females are somewhat overrepresented in the sample relative to their percentage of the general population.)

Note that this screen with the pie chart also presents other information about the variable. In the legend on the left side of the screen, you can see that the category *Male* has a 1 next to it and the category *Female* has a 2 next to it. Computers are more efficient working with numbers than with words, so researchers usually assign a number code to each category. For the moment, just ignore these number codes and focus on the category labels (male and female).

The frequency results show that 1,285 of the respondents are male and 1,619 are female. In terms of percentages, 44.2 percent are male and 55.8 percent are female. Note that the sum of these percentages will be 100 percent, although the sum will sometimes deviate slightly from 100 percent simply because of rounding errors. Now let's look at these results in a bar graph.

Data File:	**GSS96**
Task:	**Univariate**
Primary Variable:	**1) SEX**
➤ *View:*	**Bar - Freq.**

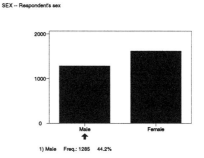

As indicated by the ➤ symbol, if you are continuing from the previous example, select the [Bar - Freq.] option (if you are using *Student MicroCase for DOS*, click [Bar] and then make sure the [Freq] option has been selected).

These results are shown in a bar graph, where each bar represents a category. The information shown below the bar graph represents the one bar that has an arrow directly beneath it. To see the values for the other bar, simply click on that bar with your mouse. Here again, we see that the bar for females is somewhat bigger than the bar for males, indicating that there are more females than males in this sample.

Let's look at another variable in the GSS96 data file.

Data File: **GSS96**
Task: **Univariate**
➤ Primary Variable: **29) WOMAN PREZ**
➤ View: **Pie**

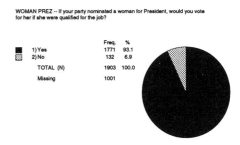

In this survey, 1,771 respondents said that they would be willing to vote for a woman for president, 132 respondents said that they would not, and there are 1,001 missing cases. In computing the percentages, the missing cases are excluded. Thus, 93.1 percent *of those who answered the question* said that they would be willing to vote for a woman for president while only 6.9 percent *of those who answered the question* said they would not be willing.

Missing data can be a problem in social research, but there actually is not a problem here with the 1,001 missing cases. In this survey, some questions were asked of only part of the sample (randomly selected). Only about two-thirds (1,960 respondents) of the sample were asked this particular question. Thus, of those who were asked this particular question, only 57 people didn't answer it.

SEARCHING FOR VARIABLES IN A DATA FILE

Return to the variable list and let's explore the Search feature. If we wanted to find a variable dealing with the religious preference of the respondent, it might take some time to scroll through the list of variables and find this particular variable. However, we can search for such a variable quite easily. This applies to any of the MicroCase tasks in which there is a list of variables.

When you have the list of variables on the screen, click on the [Search] option below the list. The box that appears will instruct you to type in a word, phrase, or character for which to search. You type in a word or phrase that might appear in either the variable name or the variable description or a variable for which you are looking.

In this example, you want to search for a variable about the religious preferences of people. Thus, you could use the term *religious preference* for a search term. Type in **religious preference** and click [OK]. As you can see, MicroCase found three variables dealing with religious preference. The first one, 18) RELIGION, is the one you want as the other two deal with the strength of religious preference.

In this particular situation, the use of the term *religious preference* narrowed the search very well. However, there are situations in which you don't want to restrict the search too much. If you wanted to find other variables in this data file about religion, you might want to search for the term *relig*. MicroCase would find any variable in the data file that contained the string *relig* anywhere in the variable name or the variable description. This procedure, for example, would select any variable that contained the word *relig*ious or the word *relig*ion.

Let's do this search. First, click on the [Full List] button to clear the results from the previous search and return to the full list of variables. Then, click the [Search] button and edit the old search term so that *relig* appears as the new search term. Click [OK] to see the results of your new search. You now see a list of ten variables that relate to religion in some way. You can view the variable descriptions from this abbreviated list of variables and select any of these variables for analysis, as you did above with the full list of variable.

After you have finished with the search option, return to the full list of variables by clicking on [Full List].

The worksheet section that follows will give you a further introduction to Student MicroCase.

NAME: _____

COURSE: _____

DATE: _____

Workbook exercises and software are copyrighted. Copying is prohibited by law.

You will need to use Student MicroCase to complete the worksheets. It is essential that you have first gone through the *Getting Started* section and the preliminary part of this exercise.

1. Let's begin by examining another data file. The ANES96 file contains selected variables from the 1996 American National Election Study. There were 1,714 respondents in this sample. Open the ANES96 data file and select the UNIVARIATE task as shown below.

> ➤ *Data File:* **ANES96**
> ➤ *Task:* **Univariate**

Scroll to variable 35) WHO VOTE? and examine the variable description. Write down the variable description and category labels below.

a. 35) WHO VOTE?
Description:

_____

_____

List the category labels:

1. _____

2. _____

3. _____

b. Now do the same thing for the variable 42) GUN LAW?.
Description:

_____

_____

List the category labels:

1. _____

2. _____

2. Use the [Search] feature to find variables in the ANES96 data file that use each of the phrases listed below. (For example, in the first situation below, you would search for any variables containing the word *abortion*.)

a. Abortion
   List the number and name of each variable found in the search.

b. Religion
   List the number and name of each variable found in the search.

c. Trust
   List the number and name of each variable found in the search.

3. a. So far, we have been looking at the variable names and descriptions. Now you will obtain univariate statistics for three variables. First, carry out the following task and fill in the category labels, the frequency in each category (the number of cases in each category), and the percentage of cases in each category.

> | | |
> |---|---|
> | *Data File:* | **ANES96** |
> | *Task:* | **Univariate** |
> | ➤ *Primary Variable:* | **55) EDUC.LEVEL** |
> | ➤ *View:* | **Pie** |

CATEGORY LABELS	FREQUENCY	PERCENT
_____	_____	_____
_____	_____	_____
_____	_____	_____
_____	_____	_____

For this variable, how many cases were missing data? _____

b. Now obtain the univariate statistics for the variable 40) SC. PRAYER, and fill in the category labels, frequencies, and percentages below.

CATEGORY LABELS	FREQUENCY	PERCENT
_____	_____	_____
_____	_____	_____
_____	_____	_____
_____	_____	_____

For this variable, how many cases were missing data? _____

c. Now obtain the univariate statistics for the variable 44) FAM VALUES and fill in the category labels, frequencies, and percentages below.

CATEGORY LABELS	FREQUENCY	PERCENT
_____	_____	_____
_____	_____	_____
_____	_____	_____

For this variable, how many cases were missing data? _____

4. Now open the USA data file so we can explore it further.

> ➤ *Data File:* **USA**
> ➤ *Task:* **Mapping**
> ➤ *Variable 1:* **33) TEEN MOMS**
> ➤ *View:* **Map**

a. Write the variable description for 33) TEEN MOMS below.

_____

_____

b. Look at the map for this variable and click on the darkest-colored states. Which region of the country has the highest rate of teenage motherhood? (Circle one.)

Northeast

Midwest

South

West

c. As indicated by this abbreviated Software Guide, use the [List: Rank] option to see the rankings of states on this variable

➤ *View:* **List: Rank**

	NAME	RATE
Highest State	_____	_____
Next Highest State	_____	_____
Third Highest State	_____	_____

d. What are the rates for the following states? (Hint: Select the option for listing cases alphabetically.)

	RATE
Florida	_____
Iowa	_____
Oregon	_____

e. Print out the ranked distribution for the TEEN MOMS variable—if you switched to the alphabetical list in the previous question, make sure to switch back to the ranked order before printing. (Note: If your computer is not connected to a printer or if you have been instructed not to use the printer, skip these printing instructions.) MicroCase will now print the ranked list for this variable.

5. For each of the following variables, map the variable, use the [List: Rank] option, and then specify the three highest and the three lowest ranked states (omit states with missing values).

a. 107) VIO.CRIME

Three Highest States	Three Lowest States
_____	_____
_____	_____
_____	_____

b. 17) % METROPOL

Three Highest States	Three Lowest States
_____	_____
_____	_____
_____	_____

6. a. Let's compare two maps.

> Data File: **USA**
> Task: **Mapping**
> ➤ Variable 1: **107) VIO.CRIME**
> ➤ Variable 2: **44) COKE USERS**
> ➤ Views: **Map**

Compare the two maps visually and then look at the rankings. Are these two maps similar to one another (darker states on one map are darker on the other, lighter states on one map are lighter on the other), opposite of one another (darker states on one map are lighter on the other, and vice versa), or neither? (Circle one.)

Similar   Opposite   Neither

b. Now compare the map of 107) VIO.CRIME with the maps for each of the variables listed below and indicate in each situation whether the two maps are similar, opposite, or neither. Circle the correct answer.

71) FLD&STREAM	Similar	Opposite	Neither
83) RENT	Similar	Opposite	Neither
108) PROP.CRIME	Similar	Opposite	Neither

7. a. Let's look at the map for the number of physicians per 100,000 population and see how this variable relates to some other characteristics of the states.

   Data File: **USA**
   Task: **Mapping**
   ➤ Variable 1: **54) MDS**
   ➤ View: **Map**

   Which region has the most states with the highest rates of physicians? (Circle one.)

   Northeast

   Midwest

   South

   West

   b. The number of physicians per 100,000 population might be directly or indirectly relate to the income level of states. Search for a variable about income and write the variable number, variable name, and variable description below.

   Variable number and name: _____

   Variable description:

   c. Compare the physician rate map with the map of the income variable you found above. Are the two maps similar to one another, opposite, or neither? (Circle one.)

   Similar   Opposite   Neither

That's all for this exercise. Remember to exit Student MicroCase properly (return to the main menu and click the [Exit] button). If you inserted your floppy diskette at the beginning of the session, remember to remove it once you have exited Student MicroCase.

Concepts and Theories

OVERVIEW

In this exercise, you will learn more about how we use concepts and theories in social research. The exercise also emphasizes the ability to recognize tautologies and the ability to identify testable statements. In addition you will learn to recognize normative statements based on values.

BEFORE YOU BEGIN

Please make sure that you have read Chapter 1 in the textbook and can answer the following review questions (you need not write any answers):

1. How are concepts and theories related to one another?

2. What is the difference between description and explanation?

3. What principles guide the construction of good concepts in social research?

4. What is a tautology?

5. What are the two essential features of theories in science?

6. Can theories be proven? Why or why not?

7. In relation to concepts, what are indicators?

8. How is the test of a hypothesis used to assess a theory?

9. What is the difference between induction and deduction?

10. What are empirical generalizations and how do they relate to induction?

11. What does it mean to say that there is a correlation between two variables?

Theoretical concepts do not stand alone, but rather are embedded in a set of relationships with other concepts. Concepts with the same name may have quite different meanings in different theories. Let's look at an example.

Sociologists have long been interested in social class, yet this concept means different things to different people. The Marxist perspective defines *social class* as the relationship of the individual to the means of production. Within this framework, there are two classes: those who own the means of production and those who don't. Other theoretical approaches divide societies into a greater number of classes, each of which can be distinguished clearly from the others.

Still other sociologists prefer to use the term *social status* since they see social status as a continuum rather than a set of identifiable classes. Even those who see social class as a continuum disagree on how it should be operationalized. Some see social class as a property of the individual, while others define it as a property of the individual's occupation. If social status, or class, is a property of the individual, then two individuals with the same occupation might have different social status. If social status is a function of the individual's occupation, then individuals with different incomes and educational levels but the same occupation will have the same social status.

None of these is the *true* definition of social class, and each definition suggests different empirical definitions. These definitions (and theories) will, however, differ in how useful they are in explaining the empirical world.

A concept should not be confused with a theory. For example, William Sims Bainbridge (1978)[1] observed an interesting phenomenon in his field study of Satan's Power, a satanic cult. When the cult first formed, members had friends outside the group as well as inside the group. As time passed, their social contacts outside the group began to decrease. This happened not only because some of the friends became members of the group, but also because members tended to stop seeing friends who were outside the cult. Eventually, members of the group had no contacts outside the group at all. Bainbridge called this phenomenon *social implosion* to describe how social relationships collapsed in upon themselves. However, naming this phenomenon added nothing to our understanding. We still don't know if this event was unique to this group or, if not, under what conditions members of groups tend to break off outside contacts or even what types of groups are likely to experience social implosion. Interesting as this phenomenon may be, we simply have a concept in need of a theory.

Confusing a concept with a theory is a relatively common problem. The difference between the two is that a theory is testable, or falsifiable, while a concept is not. We cannot say that social implosion caused this group to cut off outside contacts. Such a statement would be a tautology: Social implosion causes social implosion. On the other hand, the statement "The greater the punishment for nonconformity to group norms, the greater the likelihood that members of the group will cease contacts with outsiders" is, in principle, testable. "The more

---

[1] Bainbridge, William Sims. 1978. *Satan's Power*. Berkeley: University of California Press.

deviant the norms of the group from the norms of society, the more likely that members of the group will cease contacts with outsiders" is similarly, in principle, testable. The "in principle" caution means that we may encounter practical problems if we attempt to test the idea. For example, we may not be able to find enough appropriate groups, we may not be able to adequately measure the concept, we may encounter ethical problems, and so on.

Let's consider a second, more complicated example. Robert K. Merton (1938)[2] created the following table:

		Uses Socially Approved Means	
		Yes	No
Seeks Socially Approved Goals	Yes	Conformist	Innovator
	No	Ritualist	Retreatist

This may appear to be a theory because two different concepts are involved, socially approved means and socially approved goals. However, all we have are four different definitions. A *conformist*, for example, is defined as one who seeks socially approved goals and uses socially approved means—anyone else is not a conformist. An *innovator* is defined as one who uses means that are not socially approved to seek socially approved goals—if this is not true of the individual, then the individual is not an innovator. Each of the two remaining types are defined in a similar manner. This is called a *typology* because a combination of two characteristics is used to define a set of types.

Assuming we have information on the means used and the goals desired by a particular individual, we can place him or her in the appropriate category or type. The statement "John Dillinger was an innovator *because* he used means that were not socially approved (robbing banks) to seek goals that were socially approved (money)" is a *tautology*—it is merely the application of the definition of innovator to John Dillinger. It is not a testable statement.

If this typology were linked to some other concept, then it would be more than a set of definitions. For example, the statement "Innovators will be more common under democracies than under other forms of government" and the statement "Innovators are more likely to obtain material rewards than are conformists" are both, in principle, testable.

Now, let's take a look at a theory and how it might be tested. Most theories of crime have focused on the criminal and the characteristics that lead individuals into a life of crime. Lawrence Cohen and Marcus Felson developed an approach to deviance called *opportunity theory*. They realized that having individuals with a propensity to commit crime may be necessary but is not sufficient for a crime to occur. There also needs to be an opportunity—without banks, there can be no bank robbers. A crime requires not only a person motivated to commit the offense,

[2] Merton, Robert K. 1938. *Social Structure & Anomie*. American Sociological Review, 3:672–82.

but also the presence of a suitable target (property or individual victims) and the absence of effective guardians. For example, this theory would predict that burglaries of homes will be more likely to occur during the daytime when everyone is at school or work (the absence of effective guardians) than in the evening when homes are occupied (the presence of effective guardians).

This is a testable statement. We can check the empirical truth or falsity of this statement. If it is false, we should reject the theory.[3] If it holds true, then we can have greater faith in the theory. We can then conduct additional tests of this theory. Some empirical hypotheses provide stronger tests of theories than do others. In general, the riskier the prediction, the stronger the test. Suppose, for example, that the empirical prediction is well known to be true in advance of any research. Since we expect a theory to be consistent with known facts, this would not be a very strong test of the theory. However, if the empirical hypothesis is the opposite of what most social scientists would expect, then this would be a much stronger test of the theory.

Often a researcher may find support for a particular hypothesis and read into this additional, scientifically unjustified, conclusions. Consider a researcher who confirms the hypothesis that children who are spanked model the behavior of their parents and consequently engage in interpersonal violence. Beyond reporting this finding, the researcher adds that this study therefore proves that children should not be spanked. This is a *normative* conclusion, a statement that prescribes how people should behave based on some underlying set of values. That is, the conclusion does not logically follow from the assumptions but requires an additional value statement: Interpersonal violence is bad. Groups who value aggressive behavior might reach exactly the opposite conclusion. Normative statements cannot be evaluated in terms of true or false but only in terms of right or wrong and thus are not testable.

---

[3] In some cases, rather than reject the theory, we might reject the empirical evidence as an inadequate test of the theory. In this example, if a neighborhood has hired a security force to patrol during the daytime, then the theory would not predict a higher burglary rate during the day.

Workbook exercises and software are copyrighted. Copying is prohibited by law.

1. For each of the following statements, circle **T** if the statement is, in principle, testable or **NT** if the statement is not testable. Then, since the testability of a statement may depend on how you interpret certain elements, explain your answer in one sentence.

   a. The United States should provide low-cost health insurance to all its citizens.   T   NT

   b. Urban areas have higher crime rates than do rural areas.   T   NT

   c. People who are politically apathetic do not take much interest in politics.   T   NT

   d. The moon is made of blue cheese.   T   NT

   e. A pregnant woman whose health is endangered because of the pregnancy should be allowed to have an abortion.   T   NT

f. The United States is the most democratic country in the world.   T   NT

2. For each of the following statements, circle **Yes** if the statement is a tautology or **No** if the statement is not a tautology.

   a. Americans have more money than Brazilians because
      Americans have a higher per capita income.   Yes   No

   b. Idaho has a lower crime rate than California
      because Idaho is less urban than California.   Yes   No

   c. Mary is very popular because so many people like her.   Yes   No

   d. George votes Democratic because he has a low income.   Yes   No

   e. Susan does volunteer work because she wants to help
      other people.   Yes   No

   f. Bigamy is a crime because there are laws against it.   Yes   No

   g. Bigamy was outlawed because many people felt that it
      was morally wrong.   Yes   No

   h. Robert is a felon because he was convicted of burglary.   Yes   No

   i. Ellie has low self-esteem because her parents criticize her
      so much.   Yes   No

   j. Anyone can win an election if he or she gets enough votes.   Yes   No

3. In social research, we often use education as a variable, and we usually operationalize education as the number of years of schooling completed by an individual. If we conceptualize education as the amount of learning achieved by an individual, there are other ways that we might operationalize this concept (although these ways would not be very practical in most social research).

a. If we conceptualize education as the amount of learning achieved by an individual, what are some problems with using years of schooling to serve as an indicator of education?

b. Describe an alternate method of operationalizing education that would better reflect the amount of learning achieved.

4. The concept of *civil religion* refers to the use of religious symbols and rhetoric for public purposes and functions, especially political functions. *Civil religion* prompts U.S. presidents to mention God in the inaugural speeches. This also is the reason those in charge of designing U.S. money place "In God We Trust" on coins.

Based on materials in the textbook and the earlier discussion in this exercise, evaluate the last two sentences in the above paragraph. Does civil religion explain why U.S. presidents mention God and why "In God We Trust" is placed on coins? Discuss your answer.

5. The famous German sociologist Max Weber (1864–1920) introduced the concept of charisma to describe the situation where a leader's power is believed to rest upon divine authority—that the leader has a divine right to rule. The word itself is Greek and means "divine gift." Social scientists frequently attribute charisma to leaders. For example, because the Pope has so much charisma, many people accept his divine right to lead Christianity.

Evaluate the last sentence in this paragraph.

6. a. Give three additional empirical predictions from the opportunity theory
    of crime developed by Cohen and Felson.

        1.

        2.

        3.

    b. Assuming you have great confidence in this theory, how could you apply
    the ideas to crime prevention programs?

7. This final set of questions requires you to use Student MicroCase. (It is
    assumed that by this point you have completed the "Getting Started" sec-
    tion and the "Introductory Exercise.") Start Student MicroCase, open the
    GSS96 data file and look at the variable descriptions for variables 20) RACE
    SEG. and 30) MEN BETTER. Note that these questions are based on norma-
    tive statements and that we cannot prove or disprove these statements scien-
    tifically. However, in social research we often study the views that people
    hold on such matters.

a. Find two other variables in the GSS96 file that are based on normative statements and list them below.

   Number:_____   Name:_____

   Description:

   _____

   _____

   _____

   Number:_____   Name:_____

   Description:

   _____

   _____

   _____

b. For contrast with the normatively based questions, find two questions that are *not* based on normative statements and list them below.

   Number:_____   Name:_____

   Description:

   _____

   _____

   Number:_____   Name:_____

   Description:

   _____

   _____

2a

The Research Process Using Aggregate Data

OVERVIEW

In this exercise, you will learn more about the stages of the social research process and you will experience some of these stages directly. Specifically, you will gain experience in formulating hypotheses, selecting indicators to operationalize concepts, and using statistical results to test hypotheses. Two exercises have been designed to parallel Chapter 2 in the textbook: Exercise 2a and Exercise 2b. As the above title suggests, this first exercise focuses on the use of aggregate data. The second exercise deals with survey data. You will see the differences in the ways we utilize the two types of data as you make your way through these exercises.

BEFORE YOU BEGIN

Please make sure you have read Chapter 2 in the textbook and can answer the following review questions (you need not write any answers):

1. What are the general stages in the research process?

2. What does *operationalize a concept* mean, and how are *indicators* used in this process?

3. What is replication research and why is it important?

Many of us are concerned about violent crimes, and we wonder why some people are more likely than others to commit such crimes. Learning theorists would argue that violent criminal activities are learned in the same way that other behaviors are learned. According to this approach, to understand violent behavior, we need to see how the behavior is learned. This has led some social scientists to speculate that there should be a connection between hunting and violent crime: Hunting encourages the learning of violent behaviors, and, once learned, these behaviors can be applied to humans as well as to animals.

At this point, we have completed two steps in the research process: **select our topic** (Step 1) and **formulate our research question** (Step 2). The research question is simply "Is there a connection between hunting and violent behavior?" Next we need to **define the concepts** (Step 3):

> Hunting: the pursuit and killing of wild, game animals for sport or for food
>
> Violent behavior: physical violence, or threat of physical violence, toward another individual

Both of these concepts apply only to behaviors outside of a regular occupational role. Killing animals in an occupational role—what meat packers, chicken farmers, and humane society workers do—is not considered hunting. Further, let's exclude the threat of violence by a police officer or in a military setting from our definition of violent behavior.

When we define concepts, we also introduce the possibility that others will disagree with the definition. Because we are interested in hunting behavior, not in its motivation, we have included both hunting for food and hunting for sport in our definition. If we were testing other ideas concerned with motivation for hunting, our definition might exclude hunting for food.

We now need to **operationalize our concepts** (Step 4)—to select the indicators of these concepts. In this exercise, we'll use the **USA** data file to test this idea. The USA data set is based on aggregates, rather than individuals. That is, each case, or state, is a collection, or aggregation, of individuals.

Start Student MicroCase using the instructions in the *Getting Started* section and proceed through the tasks in the following guide—select the USA data file and select the MAPPING task.

> ➤ *Data File:* **USA**
> ➤ *Task:* **Mapping**

Based on the first exercise, we already know that we have information on hunting licenses in this data file. This could be used as our indicator of the extent of hunting in each state. Now scroll through the variable descriptions and see if you can find a variable that could be used as an indicator of violent behavior across the states. Locate variable 109) MURDER in your variable list, click on the variable to highlight it, and then read the description in the Variable Description box. This looks like a good indicator of violent behavior.

If the idea that hunting affects violent behavior is true, then we would expect states to be relatively high on both variables or to be relatively low on both variables. In other words, we would expect the maps of the two variables to look alike. We can now **formulate our hypothesis** (Step 5): States with high rates of hunting licenses will tend to have high murder rates. We are now ready to make the observations.

We are fortunate that the appropriate observations have already been collected and are included in the USA data set, so we can skip step 6, **make the observations**, and move directly to **analyze the data** (Step 7). Let's do that now.

Data File:	**USA**
Task:	**Mapping**
➤ *Variable 1:*	**72) HUNTING**
➤ *Variable 2:*	**109) MURDER**
➤ *Views:*	**Map**

HUNTING -- 1990: NUMBER OF RESIDENTS WHO PURCHASED HUNTING LICENSES PER 1,000 POPULATION (U.S.FISH & WILDLIFE)

$r = -0.466$**

MURDER -- 1992: HOMICIDES PER 100,000 POPULATION (UCR)

Remember that the ➤ symbol indicates which tasks you need to perform. You have already selected the USA data file and the MAPPING task. So here you continue by selecting the first variable. Next you select the second variable, and then you click [OK] to view the comparison of the two maps.

The map of HUNTING is shown at the top of the screen, and the map of MURDER is shown in the lower half of the screen.

The maps don't look at all alike; in fact, they appear to be almost opposites. Based on visual inspection, we would conclude that the two rates don't vary together. We have now finished Step 8—**assess the results**. (If this were an actual study, we would still have two steps remaining: publish the findings and replicate the research.)

Now let's compare the hunting licenses map with a map showing the circulation of *Field & Stream* magazine.

Data File: **USA**
Task: **Mapping**
Variable 1: **72) HUNTING**
➤ Variable 2: **71) FLD&STREAM**
➤ Views: **Map**

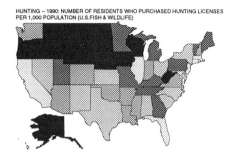

HUNTING -- 1990: NUMBER OF RESIDENTS WHO PURCHASED HUNTING LICENSES PER 1,000 POPULATION (U.S.FISH & WILDLIFE)

r = 0.829**

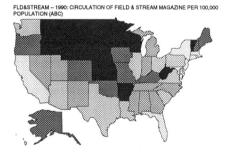

FLD&STREAM -- 1990: CIRCULATION OF FIELD & STREAM MAGAZINE PER 100,000 POPULATION (ABC)

These two maps look very much alike. However, conclusions from simply inspecting maps can be fairly subjective; our personal opinions may influence how we view the maps. So it is preferable to be able to quantify how alike or how different two maps are.

Fortunately, there is a statistical technique that will allow us to do this. So let's examine this relationship between hunting licenses and *Field & Stream* circulation differently.

Data File: **USA**
➤ Task: **Scatterplot**
➤ Dependent Variable: **71) FLD&STREAM**
➤ Independent Variable: **72) HUNTING**

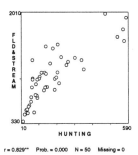

The ➤ on the Task line indicates that you must return to the main menu and select a new task—SCATTERPLOT.

A graphic appears in the middle of the screen. The y-axis is the line at the left labeled FLD&STREAM—the values on this axis range from 330 to 2010. This vertical axis represents all values of FLD&STREAM that we saw earlier on the map. The x-axis is the line at the bottom of the graph labeled HUNTING—the values on this axis range from 10 to 590. This horizontal axis represents all values of HUNTING that we also displayed earlier in a map.

Each dot on the graph represents a state and its values on these two variables. Let's see how this works. Idaho has a value of 1947 on FLD&STREAM, so it would be close to the top on the y-axis. In your mind, draw a horizontal line at approximately 1947 on the y-axis. Idaho has a value of 586.4 on HUNTING, so it would be at the far right of the x-axis. Mentally draw a vertical line at the far right of the graph. The intersection of these two lines represents Idaho.

Student MicroCase has a feature that allows you to find any case on your scatterplot. Let's use it to locate Idaho.

Data File: **USA**
Task: **Scatterplot**
Dependent Variable: **71) FLD&STREAM**
Independent Variable: **72) HUNTING**
➤ Find: **Case: Idaho**

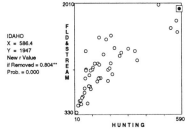

If you are continuing from the previous example, click the [Find] option. An alphabetical list of the states appears in a window. Select Idaho from the list (*Student MicroCase for DOS* requires a double-click). Then click [OK].

A highlighted square appears around the dot representing Idaho on the scatterplot. This dot is located where the two lines would intersect and represents Idaho on both variables. Further, the screen now presents the value of Idaho on each variable (listed as X and Y). You might want to find the location of some other states—click on [Find] once to deselect the current case and once more to select a new case. When you have finished, deselect the case and let's look at the regression line.

<table>
<tr><td align="right">Data File:</td><td>USA</td></tr>
<tr><td align="right">Task:</td><td>Scatterplot</td></tr>
<tr><td align="right">Dependent Variable:</td><td>71) FLD&STREAM</td></tr>
<tr><td align="right">Independent Variable:</td><td>72) HUNTING</td></tr>
<tr><td align="right">➤ View:</td><td>Reg. Line</td></tr>
</table>

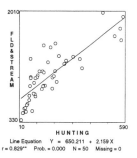

If you are continuing from the previous example, simply click the [Reg. Line] option.

A line now appears on the graph. This line represents the best effort to draw a *straight* line that connects all of the dots. It is unnecessary for you to know how to calculate the location of the regression line—the program does it for you. But if you would like to see how the regression line would look if the maps were identical, all you need to do is examine the scatterplot using the same variable for both the x-axis and the y-axis. All the dots would be on the line.

In the current example, the maps are similar but not identical, so the dots are scattered near, but not on, the regression line. Now we need a method of determining how close these dots are to the line. With the regression line showing on the scatterplot, select the [Residuals] option.

<table>
<tr><td align="right">Data File:</td><td>USA</td></tr>
<tr><td align="right">Task:</td><td>Scatterplot</td></tr>
<tr><td align="right">Dependent Variable:</td><td>71) FLD&STREAM</td></tr>
<tr><td align="right">Independent Variable:</td><td>72) HUNTING</td></tr>
<tr><td align="right">➤ View:</td><td>Reg. Line/Residuals</td></tr>
</table>

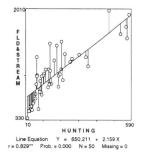

If you are continuing from the previous example, click the [Residuals] option.

A vertical line now connects each dot to the line; if we sum these distances, we can get a measure of how much alike the two maps are. The smaller this sum, the more alike are the two maps, or variables. For example, when the maps are identical and all the dots are on the regression line, the sum of these distances is zero.

This idea is used to calculate the value of a statistic called the *correlation coefficient*. The value of this coefficient can vary from −1 to +1. When the value is 1, the two maps are identical; when the value is −1, they are opposites. When the value is zero, they are neither similar nor opposites. At the bottom of the scatterplot, we can see that the correlation (indicated using a lowercase r) between hunting licenses and circulation of *Field & Stream* magazine is .829. This shows a very strong relationship between these two variables.

When the value of the correlation coefficient gets close to zero, the relationship might simply be due to chance factors, such as inaccurate measurements. How large must a correlation be to indicate that a relationship between the two variables actually exists? Using probability theory, statisticians can tell us how likely we are to observe a particular correlation coefficient by chance when there is really no relationship. Note that there are two asterisks next to the correlation ($r = .829**$). The two asterisks signify that the probability that this relationship could have occurred just because of chance factors is only 1 chance out of 100 or less. One asterisk would mean that the probability that this relationship could have occurred just because of chance factors is 5 out of 100 or less.

If this probability is small enough, then we can reject the hypothesis of no relationship, thus supporting the hypothesis that there is a relationship. In social science, the level of .05 is used for this rejection point—that is, if this correlation coefficient would be observed less than 5 times in 100 when there is no relationship, then we will conclude that a relationship probably exists. This is called the *statistical significance level*. Some researchers are even more stringent and use .01 as the rejection point. As indicated above, if the correlation reaches this level of significance (or better), then there are two asterisks next to it. You will encounter the significance level again in other types of analyses. But you will see that in some situations you will be given the exact significance level (e.g., Prob. = .013), and other times the level of significance will be indicated only by asterisks. (In Chapter 4, these issues will be discussed in greater detail.)

Let's go back to our research hypothesis about hunting and murder and look at the scatterplot for those two variables.

Data File: **USA**
Task: **Scatterplot**
➤ Dependent Variable: **109) MURDER**
➤ Independent Variable: **72) HUNTING**
➤ View: **Reg. Line**

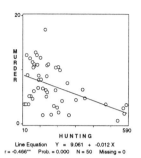

Line Equation   Y = 9.061 + -0.012 X
r = -0.466**   Prob. = 0.000   N = 50   Missing = 0

Remember, it is not necessary to return to the main menu to change your variable selections. In the Windows 95 version of the program, use the ▣ button to return to the variable selection screen (in the DOS version, simply click [Exit] once). Then, to replace variables that were previously selected, double-click on the new variables you want to use. Or, you can click the [Clear All] button and select new variables at this point.

The regression line slopes *downward* and the correlation (–0.466**) between these two variables is negative and quite strong. This relationship is significant at the .01 level—there are two asterisks next to the correlation. This means that the empirical data actually go in the opposite direction from that predicted—the higher the hunting license rate, the lower the murder rate! At this point, we would want to assess whether we really have provided a test of our research question. Perhaps we did not properly operationalize the concepts. Perhaps the observations are incorrect. There are other such problems that could have influenced the results. Some empirical observations provide much stronger tests of hypotheses than do others. By the end of the course, you should understand why this was not a particularly strong test.

However, for the present, let's assume that this was a strong test of a theory and that the empirical hypothesis was, in fact, a proper application of the theory. What can we conclude about the theory? Finding that the empirical hypothesis was not supported should lead us to reject the theory from which it was derived. The implications of the theory were wrong, therefore the theory itself must be false. On the other hand, if the empirical hypothesis were true, we could not conclude that the theory is *true*. We might have more faith in the theory, but we could not *prove* the theory. (You might want to reread the discussion of the relationship between theory and empirical observations in Chapter 1 of the text to understand this better.)

In the present example, you might find it tempting to decide that we should have used another theoretical approach. For example, you might suggest that hunting is a substitute form of violence and individuals who hunt have no need for other outlets for violent behavior. Creative social scientists could probably come up with many different ideas that would be consistent with this empirical result. However, we can't claim to have tested any of these alternatives, because we constructed them after we already knew the results of the test. Explanations of this type are called *ex post facto*—created after the fact. For research to be a valid test of a theory, the empirical hypothesis must be derived from the theory before the relevant analysis is conducted.

Incidentally, findings that fail to support a hypothesis are also important. In fact, a strong test that raises serious questions about an existing theory may change the whole course of a scientific field. Sometimes taking a step backward is really a leap forward.

The scatterplot technique we have just used is the basis for the first correlation coefficient (developed by Karl Pearson in the 1890s), and there are many variant methods based on the same underlying logic. However, it is not necessary to actually create a scatterplot in order to calculate r (the correlation coefficient), and thus it is possible to calculate many correlations at the same time. Let's use the correlation task to do this.

Data File: **USA**

➤ Task: **Correlation**

➤ Select Variables: **109) MURDER**
72) HUNTING
73) FISHING
70) PICKUPS

Correlation Coefficients
N: 50   Missing: 0
Cronbach's alpha: Not calculated--negative correlations
LISTWISE deletion (1-tailed test)   Significance Levels: ** = .01, * = .05

	MURDER	HUNTING	FISHING	PICKUPS
MURDER	1.000	-0.466 **	-0.256 *	-0.190
HUNTING	-0.466 **	1.000	0.773 **	0.709 **
FISHING	-0.256 *	0.773 **	1.000	0.609 **
PICKUPS	-0.190	0.709 **	0.609 **	1.000

The variable selection for this task works differently than other tasks you've seen. If you type in a variable name or number in the Select Variables box and press the <Enter> key, the variable name will be placed in the larger box below it. This allows you to select multiple variables for the correlation analysis. To delete a variable from this list, click on the variable name and press the <Delete> key on your keyboard (if you are using *Student MicroCase for DOS*, click the <Delete> key). Of course, you can also use the [Clear All] button to remove all variables that were previously selected.

Looking at the top left cell, we see that there is a perfect correlation (1.000) between MURDER and MURDER, as there should be since these are the same measures. Looking down the diagonal from left to right, we can see that in fact each variable is perfectly correlated with itself. Reading down the far left column, we can see the correlation between the murder rate and each of the other three variables. The correlation with HUNTING is the same as we have seen with the scatterplot. The negative signs indicate that, as each of these rates rises, the murder rate declines. Again, two asterisks indicate that a correlation is statistically significant beyond the .01 level, while one asterisk indicates significance beyond the .05 level. Notice that the correlation between MURDER and PICKUPS (the number of pickup trucks per 1,000 population) lacks an asterisk. That means it is not statistically significant and should be regarded as zero.

In addition to showing the correlation of each variable with MURDER, the matrix shows the correlations between each pair of variables. To find the correlation between any two variables, first find the name of one variable across the top of the table and then find the name of the other down the left side. Locate the cell where the two variables coincide and that is the correlation coefficient between them. Thus, for example, the correlation between HUNTING and PICKUPS is 0.709**, and between HUNTING and FISHING it is 0.773**.

In addition to using the SCATTERPLOT and CORRELATION tasks for obtaining correlations, you can see the correlation between any two aggregate variables by using the MAPPING task. Let's take a look at this.

Data File: **USA**
➤ *Task:* **Mapping**
➤ *Variable 1:* **109) MURDER**
➤ *Variable 2:* **71) FLD&STREAM**
➤ *Views:* **Map**

MURDER -- 1992: HOMICIDES PER 100,000 POPULATION (UCR)

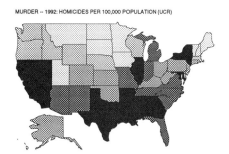

r = –0.437**

FLD&STREAM -- 1990: CIRCULATION OF FIELD & STREAM MAGAZINE PER 100,000 POPULATION (ABC)

Notice that the two maps tend to be reverse images, indicating a negative correlation. But instead of relying on visual comparisons of these two maps, look at the value of the correlation coefficient between the two variables (r = –0.437**). Whenever you compare two maps, the correlation automatically appears.

Your turn.

NAME: _____

COURSE: _____

DATE: _____

Workbook exercises and software are copyrighted. Copying is prohibited by law.

EXERCISE

2a

WORKSHEET

1. Voting for candidates is an individual matter, and most of the time we want to examine survey data in order to test hypotheses about voting choices. However, we can also discover certain voting patterns by doing aggregate analysis of election data. For example, using the USA data file, we might investigate what kind of states are more likely to vote Democratic in presidential elections and what kinds of states are more likely to vote Republican. Let's examine the map of 130) %CLINTON96.

> ➤ *Data File:* **USA**
> ➤ *Task:* **Mapping**
> ➤ *Variable 1:* **130) %CLINTON96**
> ➤ *View:* **List: Rank**

a. Look at the ten states that gave Clinton the highest percentages of their votes in the 1996 presidential election. In which region are the majority of these states located? (Circle one.)

Northeast   South   Midwest   West

b. Look at the map again to locate the states that gave the lowest percentages of their votes to Clinton. These are the light-colored states. In which region did Clinton do worst? (Circle one.)

Northeast   South   Midwest   West

c. The Democratic Party has traditionally been viewed as being more supportive of economically disadvantaged groups. Thus, we could hypothesize that Clinton would do better in states that have more poor people. Let's test this.

> *Data File:* **USA**
> ➤ *Task:* **Scatterplot**
> ➤ *Dependent Variable:* **130) %CLINTON96**
> ➤ *Independent Variable:* **86) %POOR**

What is the correlation coefficient?   r = _____

Is the correlation positive or negative?
(Circle one.)   Positive   Negative

Is the relationship statistically significant?
(Circle one.)   Yes   No

Is this hypothesis supported or rejected?
(Circle one.)   Support   Reject

d. Let's hypothesize that Clinton would do better in states that have higher unemployment.

> Data File: **USA**
> Task: **Scatterplot**
> Dependent Variable: **130) %CLINTON96**
> ➤ Independent Variable: **87) % UNEMPLOY**

What is the correlation coefficient?   r = _____

Is the correlation positive or negative?
(Circle one.)   Positive   Negative

Is the relationship statistically significant?
(Circle one.)   Yes   No

Is this hypothesis supported or rejected?
(Circle one.)   Support   Reject

e. Let's hypothesize that Clinton would do better in states that have higher percentages of their citizens receiving Aid to Families with Dependent Children (AFDC).

> Data File: **USA**
> Task: **Scatterplot**
> Dependent Variable: **130) %CLINTON96**
> ➤ Independent Variable: **79) % ON AFDC**

What is the correlation coefficient?   r = _____

Is the correlation positive or negative?
(Circle one.)   Positive   Negative

Is the relationship statistically significant?
(Circle one.)   Yes   No

Is this hypothesis supported or rejected?
(Circle one.)   Support   Reject

2. As explained in Chapter 1 of the text, evaluation research often is not prompted by theory testing. Instead, it tests empirical claims that some program or policy had its intended effect. To explore an example of such research, let's test this hypothesis: Drug education programs will reduce the incidence of drug abuse. (Note: As you select your variables for analysis, write in the complete description for each.)

> Data File: **USA**
> Task: **Scatterplot**
> ➤ Dependent Variable: **44) COKE USERS**
> ➤ Independent Variable: **45) DRUG ED**

a. Write in the description for 44) COKE USERS.

_____

_____

_____

b. Write in the description for 45) DRUG ED.

_____

_____

_____

c. What is the correlation coefficient? $r =$ _____

d. Is the correlation positive or negative?
    (Circle one.)   Positive   Negative

e. Is the relationship statistically significant?
    (Circle one.)   Yes   No

f. Is this hypothesis supported or rejected?
    (Circle one.)   Support   Reject

g. If this were the only evidence available, what conclusion would you draw about the effectiveness of drug education programs?

3. Next, test this hypothesis: School dropout rates can be reduced by increased spending on education.

Data File:	**USA**
Task:	**Scatterplot**
➤ *Dependent Variable:*	**91) DROPOUTS**
➤ *Independent Variable:*	**93) $PER PUPIL**

a. Write in the description for 91) DROPOUTS.

_____

_____

_____

b. Write in the description for 93) $PER PUPIL.

_____

_____

_____

c. What is the correlation coefficient?   r = _____

d. Is the correlation positive or negative?
   (Circle one.)   Positive   Negative

e. Is the relationship statistically significant?
   (Circle one.)   Yes   No

f. Is this hypothesis supported or rejected?
   (Circle one.)   Support   Reject

g. If this were the only evidence available, what conclusion would you draw about the effectiveness of increased school spending programs?

4. a. Aside from the murder rate, there are other ways in which we could have operationalized physical violence. Browse through the variable descriptions in the USA data file and find three other indicators of physical violence. Write the information about them below.

First indicator:
Number:_____   Name:_____
Description:

_____

_____

Second indicator:
Number:_____   Name:_____
Description:

_____

_____

Third indicator:
Number:_____   Name:_____
Description:

_____

_____

b. Now, construct a hypothesis that hunting is correlated with rates of physical violence using the first indicator you listed above.

Hypothesis 1: Hunting will be (circle choice) positively/negatively correlated with _____.
                        (first indicator)

Test this hypothesis using the SCATTERPLOT task.

Data File:	**USA**
Task:	**Scatterplot**
➤ Dependent Variable:	**(the first indicator you listed)**
➤ Independent Variable:	**72) HUNTING**

Value of r?   r = _____

Is the relationship positive or negative? (Circle one.)   Positive   Negative

Level of significance?   Prob. = _____

Is this hypothesis supported? (Circle one.)   Yes   No

c. Now construct a hypothesis that hunting is correlated with rates of physical violence using the second indicator you listed previously.

Hypothesis 2: Hunting will be (circle choice) positively/negatively correlated with _____.
                               (second indicator)

Test this hypothesis using the SCATTERPLOT task.

> Data File: **USA**
> Task: **Scatterplot**
> ➤ Dependent Variable: **(the second indicator you listed)**
> ➤ Independent Variable: **72) HUNTING**

Value of r?   r = _____

Is the relationship positive or negative? (Circle one.)   Positive   Negative

Level of significance?   Prob. = _____

Is this hypothesis supported? (Circle one.)   Yes   No

d. Now construct a hypothesis that hunting is correlated with rates of physical violence using the third indicator you listed previously.

Hypothesis 2: Hunting will be (circle choice) positively/negatively correlated with _____.
                               (third indicator)

Test this hypothesis using the SCATTERPLOT task.

> Data File: **USA**
> Task: **Scatterplot**
> ➤ Dependent Variable: **(the third indicator you listed)**
> ➤ Independent Variable: **72) HUNTING**

Value of r?   r = _____

Is the relationship positive or negative? (Circle one.)   Positive   Negative

Level of significance?   Prob. = _____

Is this hypothesis supported? (Circle one.)   Yes   No

2b

The Research Process Using Survey Data

In Exercise 2a, we worked through an example of the research process using the USA data set. Here we will examine how to study a different research question using data on individuals. In recent years, health care reform has been the subject of a heated debate. Many individuals would like the federal government to take a more active role in the administration of health care, while others prefer to leave health care in the private sector. There are many issues in this debate, but a basic disagreement is simply over the appropriate role of the federal government in programs of this type. In general, should government be more active or less active in the day-to-day life of citizens? Those with more liberal political views favor increasing the role of government, while conservatives favor limiting its role. We would expect this difference of opinion to extend to the area of health care. So we have our first research hypothesis: Those with conservative political views are less likely to favor government intervention in health care than are those with liberal political views.

We also might expect that those who are likely to be excluded under the current health care system would favor more participation by the government. We'll test this second hypothesis too.

➤ *Data File:* **GSS96**
    ➤ *Task:* **Univariate**

Let's see if we can find some variables that can be used as indicators of the relevant concepts. In the variable list, click on 59) GOV.MED. and look at its variable description:

SCALE ON GOVERNMENT COVERING MEDICAL COSTS: 1) I STRONGLY AGREE IT IS THE RESPONSIBILITY OF GOVERNMENT TO HELP; TO 5) I STRONGLY AGREE PEOPLE SHOULD TAKE CARE OF THEMSELVES.

This could be used as an indicator of whether individuals favor government participation in health care.

Now look at the variable description for 75) POL.VIEW.[1] This question could be used as an indicator of the individual's position on the liberal/conservative

---

[1] The answer categories for POL.VIEW were "recoded" by the authors so that all responses fall into three categories, rather than the original seven. You'll learn about recoding variables later in this book.

continuum. We also could use party preference, 74) POL.PARTY, and how the person voted in the 1996 presidential election, 5) WHO IN 92?, as other indicators of political conservatism.

To test the second hypothesis, we need to determine who is likely to be excluded under the current health care system. Paid health care is usually not included as a benefit for low-income jobs, so 76) INCOME might be used as an indicator of those excluded. Unemployed individuals are often uncovered as well, so 27) EVER UNEMP might be used as another indicator.

The data already have been collected so we can move to the analysis stage. First, let's look at the distribution of 59) GOV.MED.

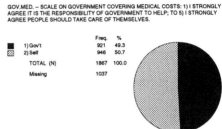

Data File: **GSS96**

Task: **Univariate**

➤ Primary Variable: **59) GOV.MED.**

➤ View: **Pie**

The results show that 49.3 percent of the sample agreed that the government should cover medical costs. We can see that there were 1,037 cases with missing data—they were either not asked the question or did not answer. Remember that missing data are excluded in all calculations, so the percentages are based only on those cases with data (1,867).

Now let's cross-tabulate views on governmental participation in health care by self-classification on the liberal-conservative dimension.

Data File: **GSS96**

➤ Task: **Cross-tabulation**

➤ Row Variable: **59) GOV.MED.**

➤ Column Variable: **75) POL. VIEW**

➤ View: **Tables**

The cross-tabulation table shows the distribution of attitude toward government participation in health care within each category of political views. For example, 290 cases indicated that they were liberal and that they supported government health care, 347 cases were moderate and supported government health care, and so on. Missing data are omitted from all calculations, so the total for the first *row* is 883 (290 + 347 + 246), and the total for the first *column* is 443 (290 + 153).

Looking at the table, we can see that, of the 443 liberals, 290 favored government health care and 153 did not. Looking at the next column, we can see that there were more moderates (347) than liberals who favored government health care but also more moderates (337) than liberals who did not. This is because more respondents in the survey identified themselves as moderates (684) than liberals (443).

This makes it obvious that we can't simply compare raw numbers of respondents. We must take differences in the size of populations into account. To do so, we can calculate the percentage of the group in each category. Since we want to compare the views on health care across the liberal-conservative political spectrum, we need to examine *column percentages*. So select the Column Percentages option.

Data File: **GSS96**
Task: **Cross-tabulation**
Row Variable: **59) GOV.MED.**
Column Variable: **75) POL. VIEW**
View: **Tables**
➤ Display: **Column %**

75) POL. VIEW

59) Gov't Self	Liberal	Moderate	Conservat.
	65.5%	50.7%	37.8%
	34.5%	49.3%	62.2%
TOTAL	100.0%	100.0%	100.0%

V=0.214**

Column percentages allow us to compare the health care views of liberals, moderates, and conservatives more easily. Looking at the first *row* of numbers and reading from left to right, we now can see that 65.5 percent of the liberals favored government health care, compared to 50.7 percent of the moderates, and only 37.8 percent of the conservatives. Thus, there is a definite pattern in the expected direction—liberals are much more favorable toward government involvement in health care than are conservatives.

While the differences are in the predicted direction, this table is based on a sample of the population, and a sample is extremely unlikely to be exactly like the population from which it is drawn. Can we be sure that this relationship actually exists in the population? We need to look at the summary statistics to assess this.

Data File: **GSS96**
Task: **Cross-tabulation**
Row Variable: **59) GOV.MED.**
Column Variable: **75) POL. VIEW**
➤ View: **Statistics (Summary)**

GOV.MED.   by   POL. VIEW

Nominal Statistics

Chi-Square: 81.271   (DF =   2; Prob. = 0.000)

V:	0.214	C:	0.209		
Lambda: (DV=75)	0.062	Lambda: (DV=59)	0.166	Lambda:	0.109

Ordinal Statistics

Gamma: s.error	0.344 0.037	Tau-b: s.error	0.201 0.022	Tau-c: s.error	0.230 0.025
Dyx s.error	0.176 0.019	Dxy: s.error	0.230 0.025		
Prob. =	0.000				

Look at the section of the screen labeled *nominal statistics*. Chi square is a statistic calculated from this table. Using probability theory, statisticians can tell us how likely we are to observe a particular chi-square value by chance when there is no relationship in the population. If this probability is small enough then we can reject the hypothesis of no relationship, which means that our hypothesis that there is a relationship is supported. In this example, the probability (Prob. = 0.000) is extremely low. Consequently, we can be pretty sure that there really is a relationship between these two variables in the population. In social science, the level of .05 is used for this rejection point—if this value of chi square would be observed less than 5 times in 100 when there is no relationship, then we will conclude that a relationship probably exists. This is called the *statistical significance level*. (In Chapter 4 of the text, these issues will be discussed in greater detail.)

We also can make use of another statistic on this screen, Cramer's V. The value of V tells us how strong the relationship is. This is analogous to Pearson's correlation coefficient (r) which we used when looking at scatterplots in Exercise 2a. V can range from 0 to 1. A value of 0 indicates no relationship between the two variables, and a value of 1 is the strongest possible relationship. In this example, V has a value of 0.214.

There's one extremely important way in which V is different than Pearson's r. Pearson's r tells us how well a straight line fits the data and also the direction of the relationship. V tells us nothing about the direction of the relationship, only about its strength. We must look at the table itself to describe the relationship. For example, if the values in the liberal and conservative column had been swapped, the value of V would be exactly the same. However, the table (with the swapped columns) would show that conservatives are more likely to support government health care and would consequently fail to support our hypothesis. When examining a relationship using tabular analysis, you must interpret the actual table, as well as examine the value and significance of V.

Let's try another test of the first hypothesis using a different indicator of the liberal-conservative continuum: political party preference.

    Data File: **GSS96**
        Task: **Cross-tabulation**
  Row Variable: **59) GOV.MED.**
➤ Column Variable: **74) POL.PARTY**
      ➤ View: **Tables**
    ➤ Display: **Column %**

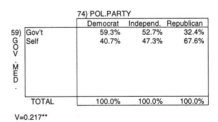

Be sure to use column percentaging on this table. In fact, nearly all cross-tabulation results in this workbook will use column percentaging. So get in the habit of selecting this option.

We see the same pattern here as well: Democrats are more likely than independents, and independents are more likely than Republicans, to endorse government health care. Let's check the summary statistics.

Data File: **GSS96**

Task: **Cross-tabulation**

Row Variable: **59) GOV.MED.**

Column Variable: **74) POL.PARTY**

➤ View: **Statistics (Summary)**

GOV.MED   by   POL.PARTY

Nominal Statistics

Chi-Square: 85.981	(DF =	2; Prob. = 0.000)			
V:	0.217	C:	0.212		
Lambda:	0.018	Lambda:	0.170	Lambda:	0.086
(DV=74)		(DV=59)			

Ordinal Statistics

Gamma:	0.328	Tau-b:	0.193	Tau-c:	0.222
s.error	0.036	s.error	0.021	s.error	0.025
Dyx	0.168	Dxy:	0.222		
s.error	0.019	s.error	0.025		
Prob. =	0.000				

This relationship is statistically significant at the .000 level, and V is .217. Let's test our hypothesis yet another time using another indicator (presidential voting preference) of the liberal-conservative continuum.

Data File: **GSS96**

Task: **Cross-tabulation**

Row Variable: **59) GOV.MED.**

➤ Column Variable: **5) WHO IN 92?**

➤ View: **Tables**

➤ Display: **Column %**

5) WHO IN 92?

59) Gov't	Clinton	Bush	Perot
Self	58.3%	29.9%	46.8%
	41.7%	70.1%	53.2%
TOTAL	100.0%	100.0%	100.0%

V=0.259**

We can see that those who voted for Clinton were more likely to favor government health care than those who voted for Bush or Perot. Those who voted for Perot were more likely to favor government health care than those who voted for Bush. Check the statistics for this relationship and you will see that it is statistically significant at the .000 level, and V is .259.

Since our hypothesis really should be limited to the Democratic and Republican candidates, we should have eliminated the Perot voters from this analysis. We can accomplish this several different ways, but the easiest method is to use the [Collapse] option that is provided with all cross-tabulation results. This feature allows us to drop the Perot supporters from the present analysis. First, with the cross-tabulation table on the screen, click on the label Perot in the Perot column. Note that this highlights the column. Now click on the [Collapse] button. In response to the question, click on the option which allows you to *drop* the category—convert it to missing data for the present analysis. Click [OK] to view the updated table. Note that this has no effect at all on the column percentages for the Clinton and Bush columns. However, this procedure might change the significance level and Cramer's V. In this case, the significance level is still .000 and V is slightly higher (.282) than it was before dropping the Perot supporters from the analysis.

	Data File:	**GSS96**
	Task:	**Cross-tabulation**
	Row Variable:	**59) GOV.MED.**
	Column Variable:	**5) WHO IN 92?**
	View:	**Tables**
	Display:	**Column %**

59) GOV.MED.	5) WHO IN 92?	
	Clinton	Bush
Gov't	58.3%	29.9%
Self	41.7%	70.1%
TOTAL	100.0%	100.0%

V=0.282**

All three tests of the first research hypothesis (that liberals are more support-ive of government participation in health care than are conservatives) provided support for it. Let's turn to our second hypothesis—the expectation that those who are likely to be excluded under the current health care system would favor more participation by the government.

	Data File:	**GSS96**
	Task:	**Cross-tabulation**
	Row Variable:	**59) GOV.MED.**
➤	Column Variable:	**76) INCOME**
➤	View:	**Tables**
➤	Display:	**Column %**

59) GOV.MED.	76) INCOME		
	Under 15K	15K-29,999	30K & over
Gov't	57.9%	52.3%	46.2%
Self	42.1%	47.7%	53.8%
TOTAL	100.0%	100.0%	100.0%

V=0.095**

Those in the lowest income category are the most supportive of government participation in health care, those in the next income category are somewhat less supportive, and those in the highest income category are the least supportive. The differences are not especially large so let's check the summary statistics for this relationship. As we see, the relationship is statistically significant (Prob. = .000) and V is .095. Thus, there is a connection between income and views on govern-ment participation in health care although the relationship is not very strong. Let's try another test of this hypothesis using the question of whether the respon-dent has ever been unemployed.

Data File: **GSS96**
Task: **Cross-tabulation**
Row Variable: **59) GOV.MED.**
➤ Column Variable: **27) EVER UNEMP**
➤ View: **Tables**
➤ Display: **Column %**

59) GOV. MED	27) EVER UNEMP	
	Yes	No
Gov't	57.6%	45.2%
Self	42.4%	54.8%
TOTAL	100.0%	100.0%

V=0.117**

The results are in the predicted direction—those who have been unemployed at some point are more favorable toward government participation in health care. When we check the statistics for this relationship, we see that it is statistically significant (Prob. = .000) and V = .117. Thus, both tests support our second hypothesis that those who are likely to be excluded under the current health care system would favor more participation by the government.

We found fairly strong support for our first hypothesis and moderate support for our second hypothesis.

Your turn.

NAME: _____

COURSE: _____

DATE: _____

Workbook exercises and software are copyrighted. Copying is prohibited by law.

EXERCISE

2b

WORKSHEET

1. Let's use 41) FEAR CRIME from the ANES96 data file to explore the research question "What kinds of people are most likely to fear crime?" First, we need to see whether there are differences among people in the extent to which they fear crime.

    > *Data File:* **ANES96**
    > *Task:* **Univariate**
    > *Primary Variable:* **41) FEAR CRIME**
    > *View:* **Pie**

    a. Fill in the percentages below.

	Frequency	Percent
Very afraid	_____	_____ %
Somewhat afraid	_____	_____ %
Little bit afraid	_____	_____ %
Not afraid	_____	_____ %
Total	_____	_____ %

    Thus not everyone is equally afraid of crime. What kinds of people are more likely to fear crime? People who are more vulnerable to crime might fear crime more. Thus, we might expect older people to fear crime more than younger people, females to fear crime more than males, and African-American citizens to fear crime more than white citizens. Let's test each of these propositions.

    b. Do the following analysis and fill in the information in the table. Use column percentages for the table.

    > *Data File:* **ANES96**
    > *Task:* **Cross-tabulation**
    > *Row Variable:* **41) FEAR CRIME**
    > *Column Variable:* **54) AGE CATEGR**
    > *View:* **Tables**
    > *Display:* **Column %**

    If the entire table does not fit on your screen, use the scroll bars.

	Under 30	30–44	45–59	60 and up
Very afraid	_____%	_____%	_____%	_____%
Somewhat afraid	_____%	_____%	_____%	_____%
Little bit afraid	_____%	_____%	_____%	_____%
Not afraid	_____%	_____%	_____%	_____%

V = _____

Prob. = _____

Based on the results above, would you conclude that
older people fear crime more than younger people do?
(Circle one.)   Yes   No

c. Now let's see if there is a gender difference when it comes to fear of
crime.

Data File: **ANES96**
Task: **Cross-tabulation**
Row Variable: **41) FEAR CRIME**
➤ Column Variable: **58) SEX**
➤ View: **Tables**
➤ Display: **Column %**

Fill in the information in the following table for this analysis. Use column
percentages for the table.

	Male	Female
Very afraid	_____%	_____%
Somewhat afraid	_____%	_____%
Little bit afraid	_____%	_____%
Not afraid	_____%	_____%

V = _____

Prob. = _____

Based on the results above, would you conclude that
females fear crime more than males do? (Circle one.)   Yes   No

d. Now compare African Americans and white Americans.

> Data File: **ANES96**
> Task: **Cross-tabulation**
> Row Variable: **41) FEAR CRIME**
> ➤ Column Variable: **59) RACE**
> ➤ View: **Tables**
> ➤ Display: **Column %**

Since we want to compare white Americans with African Americans, we need to remove Native Americans and Asian Americans from the analysis. First, click on the label NAT. AMER in the table to highlight the column for Native Americans. Next, click on the label ASIAN in the table to highlight the Asian Americans column. Then click the [Collapse] button and convert these cases to missing data (drop them from the present analysis). Click [OK] to return to the table. Now fill in the information below.

	White	Black
Very afraid	_____%	_____%
Somewhat afraid	_____%	_____%
Little bit afraid	_____%	_____%
Not afraid	_____%	_____%

V = _____

Prob. = _____

Based on the results above, would you conclude that African Americans fear crime more than white Americans do? (Circle one.)   Yes   No

2. Let's return to the GSS96 data file and look at another relationship concerning attitudes toward government involvement in health care.

> ➤ Data File: **GSS96**
> ➤ Task: **Cross-tabulation**

a. First, fill in the description of 67) OVER 50.

_____

_____

b. Follow the instructions below and fill in the column percentages in the table.

Data File:	**GSS96**
Task:	**Cross-tabulation**
➤ Row Variable:	**59) GOV.MED.**
➤ Column Variable:	**67) OVER 50**
➤ View:	**Tables**
➤ Display:	**Column %**

	Under 50	50 & over
Gov't.	_____%	_____%
Self	_____%	_____%

Describe the relationship (the pattern in the results) in this table.

c. Now check the summary statistics for this relationship.

Data File:	**GSS96**
Task:	**Cross-tabulation**
Row Variable:	**59) GOV.MED.**
Column Variable:	**67) OVER 50**
➤ View:	**Statistics (Summary)**

V = _____

d. Is the relationship statistically significant at the .05 level or better? (Circle one.)   Yes   No

3. The variable 60) MUCH GOV'T directly taps opinions about government involvement in day-to-day life.

Data File:	**GSS96**
Task:	**Cross-tabulation**

a. Write the description of 60) MUCH GOV'T.

_____

_____

_____

_____

b. Using the same indicators of political conservatism used previously in the exercise, test the hypothesis that conservatives are more opposed to government involvement in day-to-day life. Present the results in the following tables. (Remember to use column percentaging.)

➤ *Row Variable:* **60) MUCH GOV'T**
➤ *Column Variable:* **75) POL. VIEW**

	Liberal	Moderate	Conservative
Too little	_____%	_____%	_____%
Both/2 much	_____%	_____%	_____%

V = _____

Prob. = _____

Row Variable: **60) MUCH GOV'T**
➤ *Column Variable:* **5) WHO IN 92?**

	Clinton	Bush	Perot
Too little	_____%	_____%	_____%
Both/2 much	_____%	_____%	_____%

V = _____

Prob. = _____

Row Variable: **60) MUCH GOV'T**
➤ *Column Variable:* **74) POL.PARTY**

	Democrat	Independent	Republican
Too little	_____%	_____%	_____%
Both/2 much	_____%	_____%	_____%

V = _____

Prob. = _____

    c. In a few sentences, summarize the results of this analysis. To what extent are these indicators related to people's attitudes toward the role of government?

4.   You have opened a market research firm and your first job is to help your local art museum plan a campaign to increase attendance. The museum directors have received a donation that is to be spent on radio advertising. Your job is to select an all-music station on which to run the museum commercials. We will use the GSS93 data set for this research.

    ➤ *Data File:* **GSS93**
    ➤ *Task:* **Cross-tabulation**

Look at the variable descriptions for variables 11 to 21. These tell us about individuals' musical preferences. Variable 23) VISIT ART tells us how often individuals visit art museums.

Select two types of music you think will be *positively* related to visiting museums; that is, the more the person likes the music, the *more likely* they are to visit museums. Fill in the following information for each of these two variables.

    a. Row variable: 23) VISIT ART
       Column variable: _____

	Very much	Like	Mixed	Dislike	Very Much
Yes	_____%	_____%	_____%	_____%	_____%
No	_____%	_____%	_____%	_____%	_____%

V = _____

Prob. = _____

Are people who like this type of music more or less
likely to visit art museums than are people who
dislike this type of music? (Circle one.)   More Likely

                                                             Less Likely

                                                             No Difference

b. Row variable: 23) VISIT ART
   Column variable: _____

	Very much	Like	Mixed	Dislike	Very much
Yes	_____%	_____%	_____%	_____%	_____%
No	_____%	_____%	_____%	_____%	_____%

                                                             V = _____

                                                             Prob. = _____

Are people who like this type of music more or less
likely to visit art museums than are people who
dislike this type of music? (Circle one.)   More Likely

                                                             Less Likely

                                                             No Difference

5. Now, select two types of music you think will be *negatively* related to visiting museums; that is, the more the person likes the music, the *less likely*
   they are to visit museums. Fill in the following information for each of
   these two variables.

a. Row variable: 23) VISIT ART
   Column variable: _____

	Very much	Like	Mixed	Dislike	Very much
Yes	_____%	_____%	_____%	_____%	_____%
No	_____%	_____%	_____%	_____%	_____%

                                                             V = _____

                                                             Prob. = _____

Are people who like this type of music more or less
likely to visit art museums than are people who
dislike this type of music? (Circle one.)   More Likely

Less Likely

No Difference

b. Row variable: 23) VISIT ART
   Column variable: _____

	Very much	Like	Mixed	Dislike	Very much
Yes	_____%	_____%	_____%	_____%	_____%
No	_____%	_____%	_____%	_____%	_____%

$V = $ _____

Prob. = _____

Are people who like this type of music more or less
likely to visit art museums than are people who
dislike this type of music? (Circle one.)   More Likely

Less Likely

No Difference

c. Based on the tables in Questions 4 and 5, what recommendations would
   you make to the local art museum? Provide statistical evidence to support
   your answer.

Evaluating Indicators

OVERVIEW

In this exercise, you'll learn more about measurement in social research. This will include further experience with different units of analysis (cases) and different levels of measurement. We also will delve more into reliability and validity, and see why we place more confidence in the accuracy of relationships among variables than we do in the accuracy of our measurements of the variables themselves.

BEFORE YOU BEGIN

Please make sure you have read Chapter 3 in the textbook and can answer the following review questions (you need not write any answers):

1. What does it mean to say that variables have variation within them?

2. Describe the levels of measurement: nominal, ordinal, interval, and ratio.

3. What are units of analysis, and what are some of the different kinds of units of analysis used in social research?

4. What are aggregate data, and how might misinterpretation of results based on aggregate data lead to the ecological fallacy?

5. What is reliability, and what methods can we use to assess the degree of reliability of our measurements of variables?

6. What is validity, and what are the common methods of assessing validity?

7. Compare and contrast indexes (or indices) and scales.

8. When using aggregate data, why is it often important to convert raw numbers (e.g., number of murders by state) into rates (e.g., murders per 100,000 population by state)?

In some of the physical sciences, measurement techniques are very sophisticated and precise because theories dictate exactly how various concepts should be measured. In the social sciences, we are not so fortunate, and a concept legitimately can be measured in many different ways. Consequently, our measures are much cruder and many ambiguities are introduced into the research process. In this exercise, we'll look at some of the special problems of measurement in social science.

UNITS OF ANALYSIS

Not all concepts used in the social sciences apply to individuals. Racial segregation, for example, refers to the extent to which two or more racial groups are separated from each other and, therefore, cannot be a property of individuals but only of *groups* of individuals. You can measure the racial segregation of cities, churches, schools, and so on, but you cannot measure the racial segregation of individuals.

Ideally, the unit of analysis used in the research hypothesis should be the same unit used in the research question. However, with some research questions, there are practical problems. For example, suicide is a relatively rare event. And since victims of suicide are not available to answer questions, studies of suicide frequently use aggregate units to test ideas.

In Exercise 2a, we were interested in the research question that hunting is a cause of violent behavior. We then looked at the relationship between the hunting license rate and indicators of violent behaviors using states as the unit of analysis. Notice that the research question deals with the individual as the unit of analysis: Are individuals who hunt more likely to behave violently than individuals who don't hunt? However, the research hypothesis uses states as the unit of analysis and looks at the connection between the hunting license rate and the murder rate. Even if the analysis at the state level had supported the research hypothesis, there would be no direct support for the underlying research question, which is trying to assess individual behavior. To do so would be to commit the ecological fallacy. When we found that the hunting license rate and the murder rates were negatively correlated, we could not necessarily conclude that they were also negatively correlated at the individual level. Similarly, we could not conclude that variables correlated at the individual level necessarily would be correlated at the aggregate level.

But even studies based on less than ideal units of analysis can contribute to knowledge. For example, based on the strong negative relationship between the hunting license rate and the murder rate that we observed, we can probably safely conclude that hunters are not *more* likely to commit murder.

AGGREGATES AND RATES

Sometimes, units or cases are combined to create an entirely new unit of analysis. Let's say, for example, that we wanted to examine the effect of the degree of school segregation on racial prejudice of students. One way to proceed with such a study would be to conduct surveys of students at schools that differed greatly in their racial composition, examine prejudice within those schools, and then compare the results across schools. For the first stage of the study (i.e., examining prejudice within schools), students would be the unit of analysis. But, when comparisons across the schools were made, schools would be the unit of analysis.

When aggregating information across individuals to obtain an aggregate measure, you always should transform the number into a rate for use in analysis. Let's take a look at what is meant by this.

➤ *Data File:* **USA**
  ➤ *Task:* **Mapping**
➤ *Variable 1:* **59) #PLAYBOY**
➤ *Variable 2:* **2) POP 1990**
  ➤ *Views:* **Map**
  ➤ *Display:* **Spot Fill**

#PLAYBOY -- 1990: CIRCULATION OF PLAYBOY MAGAZINE (IN 1000S)

r = 0.991**

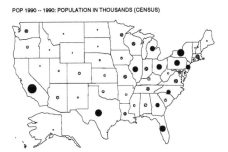

POP 1990 -- 1990: POPULATION IN THOUSANDS (CENSUS)

The two maps are virtually identical. All we are seeing is that *Playboy* sells more copies in states with more people, which is hardly surprising. What we really want to know is which states have the highest *rates* of readership of *Playboy*. Hence, in order to make meaningful comparisons, we need to standardize across states on the basis of population.

Data File: **USA**
Task: **Mapping**
Variable 1: **59) #PLAYBOY**
➤ *Variable 2:* **58) PLAYBOY**
➤ *Views:* **Map**
➤ *Display:* **Spot Fill**

#PLAYBOY -- 1990: CIRCULATION OF PLAYBOY MAGAZINE (IN 1000S)

r = –0.142

PLAYBOY -- 1990: PLAYBOY CIRCULATION PER 100,000 POPULATION (ABC)

The map for PLAYBOY shows the circulation of *Playboy*, divided by the population of the state and then multiplied by 100,000. Thus, this map shows the number of *Playboy* copies sold per 100,000 population. This map is very different from the map of raw circulation numbers.

In summary, a rate is created by reducing the numbers for each unit—in this situation, each state—to a common base. Social researchers often use population as their common base, as was done in this example. This places populous states, such as California and Texas, on equal footing with states such as Wyoming and North Dakota.

LEVEL OF MEASUREMENT

Level of measurement is sometimes explained as the extent to which the categories of the variable reflect the properties of the number system. In the case of *nominal variables*, we can only sort cases into groups, such as males and females. Even if we assign a number to each group (0 for male and 1 for female), these numbers simply tell us which cases have the same gender and which cases have different genders. With *ordinal variables*, we can not only sort the cases into groups, but also order cases in terms of the property to be measured—the higher the number the more of the property possessed by the case. *Interval variables* have yet another property of true numbers—there is an equal distance between categories. *Ratio variables* not only have equal distances between categories, but also have a rational zero point. If a case is zero on a ratio variable, this case has *none* of the property being measured.

In daily life, almost all numbers we encounter (except identification numbers such as your social security number) have all of the properties of true numbers—having zero in our checking account has a very definite meaning. Unfortunately, in social science, many of the numbers we use are not really numbers—they lack one or more of the necessary properties. Therefore, we are restricted in how we can work with them. This is why the level of measurement of a variable is important. The level reflects which properties are possessed by the variable.

The numbers we assign to categories of variables, referred to as *codes*, are necessary to do quantitative social science research. But just because we are assigning numbers doesn't mean the categories actually reflect all of the properties of these numbers. If we are using a race/ethnicity variable that is coded 0 for category "white," 1 for category "African American," 2 for "Asian American," and so on, we cannot use these numbers to order cases from less to more on our race variable. We cannot say, for example, that zero represents the absence of race. We are merely using numbers to show whether cases are in the same category or in different categories.

When we analyze data, we use many different techniques, or statistics, for summarizing the observed data. Each of these techniques makes assumptions about the numeric properties possessed by the variable (i.e., its level of measurement). A frequency distribution is the simplest type of summary statistic and can be used with any level of measurement. A frequency distribution is the simplest type of summary statistic and can be used with any level of measurement. Let's examine the frequency distribution for a nominal variable.

➤ *Data File:* **GSS96**

➤ *Task:* **Univariate**

➤ *Primary Variable:* **18) RELIGION**

➤ *View:* **Statistics (Summary)**

RELIGION -- What is your religious preference? Is it Protestant, Catholic, Jewish, some other religion, or no religion?

Mean: 1.831   Std.Dev.:   1.217   N: 2899
Median: 1.000   Variance: 1.481   Missing: 5
99% confidence interval +/- mean: 1.773 to 1.890
95% confidence interval +/- mean: 1.787 to 1.876

	Freq.	%	Cum.%	Z-Score
1) Protestant	1664	57.4	57.4	-0.683
2) Catholic	685	23.6	81.0	0.139
3) Jewish	68	2.3	83.4	0.960
4) None	339	11.7	95.1	1.782
5) Other	143	4.9	100.0	2.604

Notice that numbers have been assigned to the categories: Protestants are coded 1, Catholics are coded 2, and so on. This is a nominal variable, since cases can only be categorized, not ordered, by religious affiliation.

The category Protestant has more cases than any other category. Again, do not confuse the category *code*, 1, with the number of cases in the category, 1664. Now let's examine the frequency distribution for an ordinal variable.

Data File: **GSS96**

Task: **Univariate**

➤ Primary Variable: **99) CH.ATTEND!**

➤ View: **Statistics (Summary)**

CH.ATTEND! -- How often do you attend religious services?

Mean:	3.702	Std.Dev.:	2.641	N:	2823
Median:	3.000	Variance:	6.973	Missing:	81

99% confidence interval +/- mean: 3.574 to 3.830
95% confidence interval +/- mean: 3.604 to 3.799

	Freq.	%	Cum.%	Z-Score
0) Never	438	15.5	15.5	-1.402
1) < 1 a year	253	9.0	24.5	-1.023
2) 1 or 2 yr.	400	14.2	38.6	-0.644
3) Sev a year	416	14.7	53.4	-0.266
4) 1 a month	190	6.7	60.1	0.113
5) 2-3 month	271	9.6	69.7	0.492
6) About wkly	160	5.7	75.4	0.870
7) Weekly	486	17.2	92.6	1.249
8) Sev. a wk.	209	7.4	100.0	1.628

The categories of CH.ATTEND! are also coded with numbers: Those who never attend church are coded 0, those who attend only once a year or less often are coded 1, and so on. These codes allow us not only to group cases as similar or different, but also to *order* cases. So this variable is an ordinal variable. By convention, the lowest category is usually coded 1 (or 0), and the code is incremented by 1 for each additional category.

Look at the column labeled "Cum. %." The numbers in this column represent the percent of cases at or below that category. For example, we see that 60.1 percent attend church once a month or less often. If we couldn't order the categories, this summary measure wouldn't be valid. Looking at the previous example, we could not have said that 86.2 percent of the sample are Catholic or lower—no one would know how to interpret the statement. So cumulative percents should be used only with ordinal or higher levels of measurement.

Percentile is another word for cumulative percentage. The 90th percentile of a group on a particular variable is the category at or below which 90 percent of the group scored. The *median* is the 50th percentile. The median is a commonly used measure because it splits a sample in half—50 percent of the sample are at or below the median value. Next, let's look at a ratio variable.

Data File: **GSS96**

Task: **Univariate**

➤ Primary Variable: **90) # SIBS!**

➤ View: **Statistics (Summary)**

SIBS! -- How many brothers and sisters did you have? Please count those born alive, but no longer living, as well as those alive now. Also include stepbrothers and stepsisters, and children adopted by your parents.

Mean:	3.864	Std.Dev.:	3.522	N:	2897
Median:	3.000	Variance:	12.407	Missing:	7

99% confidence interval +/- mean: 3.695 to 4.033
95% confidence interval +/- mean: 3.736 to 3.992

	Freq.	%	Cum.%	Z-Score
0	140	4.8	4.8	-1.097
1	501	17.3	22.1	-0.813
2	570	19.7	41.8	-0.529
3	497	17.2	59.0	-0.245
4	350	12.1	71.0	0.039
5	228	7.9	78.9	0.323
6	154	5.3	84.2	0.606
7	136	4.7	88.9	0.890
8	84	2.9	91.8	1.174
9	53	1.8	93.6	1.458
10	47	1.6	95.3	1.742

Since the categories of # SIBS! are ordered, the median is meaningful. In the statistics listed on the screen, you can see that the median number of siblings for this sample was 3.00. This means that half of the sample had 3 sibs or fewer and the other half had more.

Because this measure possesses equal distance between categories (a requirement of interval and ratio variables), we can add category values. Consequently, we can find the *average*, or *mean*, value of the variable. The average is found by adding the category values over all cases and dividing by the number of cases. Looking at the statistics on the screen, we can see that the average number of siblings for this sample was 3.864.

Both the median and the mean tell us something about the location of the middle of the distribution. Such statistics are broadly referred to as *measures of central tendency*.

Statistics that summarize the relationship *between* variables also make assumptions about the level of measurement of the variables involved. For example, in Exercise 2b, we saw that Cramer's V provides information on a cross-tabulation that is similar to the information provided by Pearson's r on a scatterplot. The main difference between these statistics and their interpretation is the level of measurement of the variables. Scatterplots and Pearson's r assume that the variables are measured at the interval or ratio level, while Cramer's V only requires nominal level of measurement. Let's look at the statistics for a cross-tabulation table.

Data File:	**GSS96**	
➤ *Task:*	**Cross-tabulation**	
➤ *Row Variable:*	**70) PRAY**	
➤ *Column Variable:*	**64) DEGREE**	
➤ *View:*	**Tables**	
➤ *Display:*	**Column %**	

64) DEGREE

70) PRAY	Not hi sch	High sch	Some coll.
Daily	67.6%	57.8%	53.2%
Less	32.4%	42.2%	46.8%
TOTAL	100.0%	100.0%	100.0%

V=0.091*

Row percents, column percents, and total percents can be used with nominal variables. We'll use only column percents in the exercises. Notice that there is a pattern in these result—there are differences in the frequency of praying among those who didn't finish high school, high school graduates, and those with some college. Now we need to examine the statistics for this relationship between frequency of praying and degree.

Data File: **GSS96**
Task: **Cross-tabulation**
Row Variable: **70) PRAY**
Column Variable: **64) DEGREE**
➤ View: **Statistics (Summary)**

PRAY   by   DEGREE

Nominal Statistics

Chi-Square: 8.155 (DF = 2; Prob. = 0.017)					
V:	0.091	C:	0.091		
Lambda:	0.000	Lambda:	0.000	Lambda:	0.000
(DV=64)		(DV=70)			

Ordinal Statistics

Gamma:	0.152	Tau-b:	0.081	Tau-c:	0.087
s.error	0.057	s.error	0.030	s.error	0.032
Dyx:	0.074	Dxy:	0.089		
s.error	0.028	s.error	0.033		
Prob. =	0.007				

Notice the "Nominal Statistics" heading at the top of the screen. All the statistics in this group are appropriate for use with nominal variables. Remember that, while Cramer's V indicates the *strength* of the relationship, it does not tell us the direction or nature of the relationship. To describe the relationship, you must look at the table itself.

In the middle of the screen, you will see another set of statistics (gamma, tau-b, tau-c, dyx, and dxy) that are for ordinal variables. Since ordinal variables (and variables with higher levels of measurement) have a natural direction or order to their categories, the relationship between variables can be either positive or negative—as one variable increases, the other variable may either increase or decrease. Hence, the statistics under this heading describe both the strength and the direction of the relationship, just as the correlation coefficient does.

Your instructor may wish you to report the value and significance of gamma when both variables are ordinal. Gamma may range from –1 through 0 to +1, where the sign shows the direction of the relationship and the value indicates the strength. The significance of all the ordinal statistics, including gamma, is shown at the bottom (Prob. = 0.007).

Be alert to the fact that MicroCase will provide values for all the statistics regardless of the level of measurement—this is true of all leading statistical packages. Consequently, in any analysis *you* must determine which statistics are appropriate. For example, it would be inappropriate to report the value of gamma for nominal variables.

As pointed out in the text, few social science variables based on individuals can be considered truly interval or ratio. For example, while years served in prison may be a ratio measure of severity of sentence, years spent in school is probably not a ratio measure of education. Unfortunately, the most powerful statistical techniques assume at least an interval level of measurement. But the use of these techniques on ordinal-level variables does not seem to generate misleading conclusions in most applications, and statisticians are even finding ways to use these techniques with variables measured at the nominal level.

In this course, unless your instructor requests otherwise, we will assume that correlation and other related techniques can be used with ordinal or higher levels of measurement. You'll learn more about potential problems with this assumption when you take a statistics course.

RELIABILITY AND VALIDITY

Based on the discussion in the text, you may have noticed that we always assess reliability by looking at the relationship *between* two variables:

- Inter-rater reliability—relationship between ratings by different individuals
- Test-retest reliability—relationship between scores on the same test at different times
- Alternate forms reliability—relationship between scores on two different forms of the same test
- Split halves reliability—relationship between scores on two different forms administered at the same time
- Internal consistency reliability—relationships among items used in an index

Let's look at Cronbach's alpha—a measure of reliability based on internal consistency.

> *Data File:* **GSS96**
> ➤ *Task:* **Correlation**

First, look at the variable description for 52) FREE SPEAK. This support for freedom of speech index is based on answers to five questions (variables 47–51) asking whether the respondent would allow a public speech by a person opposed to religion and churches, a racist, a communist, a militarist, and a homosexual.

Is this measure reliable in terms of internal consistency? It is possible that some of these questions are reliable indicators while others are not. While social scientists usually focus more on validity than on reliability, let's look at one common measure of reliability that could be used here: Cronbach's alpha coefficient. Alpha is used to measure internal consistency—based on the extent to which the items are correlated with one another—especially when a researcher is developing an index to measure some concept.

Cronbach's alpha varies from 0 (completely unreliable) to 1.0 (perfectly reliable in terms of internal consistency). If alpha is 0.7 or higher, then the index is considered to be reliable. Let's check the reliability for the support for freedom of speech index.

> *Data File:* **GSS96**
> *Task:* **Correlation**
> ➤ *Select Variables:* **47) ATHEIST SP**
> **48) RACIST SPK**
> **49) COMMUN SPK**
> **50) MILITI. SP**
> **51) GAY SPEAK**

Correlation Coefficients
N: 1779   Missing: 1125
Cronbach's alpha: 0.818
LISTWISE deletion (1-tailed test)   Significance Levels: ** = .01, * = .05

	ATHEIST SP	RACIST SPK	COMMUN SPK	MILITI. SP	GAY SPEAK
ATHEIST SP	1.000	0.513 **	0.563 **	0.496 **	0.472 **
RACIST SPK	0.513 **	1.000	0.490 **	0.479 **	0.355 **
COMMUN SPK	0.563 **	0.490 **	1.000	0.562 **	0.427 **
MILITI. SP	0.496 **	0.479 **	0.562 **	1.000	0.391 **
GAY SPEAK	0.472 **	0.355 **	0.427 **	0.391 **	1.000

At the top of the screen, you can see that Cronbach's alpha is 0.818 and hence our index is reliable. You can see that this high value results from the high correlations among the component variables.

Similarly, all tests of validity, except face validity, are also based on examining the relationship between variables. For example, in the GSS96 data set, individuals are asked how often they go to church and also how strong their religious preference is. If the measure of religiosity is valid, we would expect that those who go to church more often would have stronger religious preferences. So we could look at the relationship between these variables to test the validity of the religiosity measure.

		72) CH.ATTEND	
		Not often	Often
71) HOW RELIG?	Strong	20.5%	67.1%
	Less	79.5%	32.9%
	TOTAL	100.0%	100.0%

Data File: **GSS96**
➤ Task: **Cross-tabulation**
➤ Row Variable: **71) HOW RELIG?**
➤ Column Variable: **72) CH.ATTEND**
➤ View: **Tables**
➤ Display: **Column %**

V=0.470**

We can see from these column percentages that those who attend church often claim to have stronger religious preferences than those who don't. (Further, a check of the statistics shows that Cramer's V = .470 and Prob. = .000.) Hence, we have more confidence in the validity of our measure of religiosity.

WHY RELATIONSHIPS?

This focus on relationships is also a result of the crudeness of our measures. Very few social science variables have categories that allow us to give meaningful interpretations to the distribution of a single variable. Generally, only variables with "natural" categories, such as sex, race, and political party affiliation, or variables measured at the ratio level, such as age and income, provide meaningful distributions. For example, the statement that a group is 70 percent female or the statement that the average age of the group is 63.5 can be unambiguously interpreted. However, distributions of single variables that have a natural continuum but no meaningful zero point, such as degree of belief in God, attitude toward abortion, or degree of confidence in the government, cannot be interpreted easily.

For example, the statement that 60 percent of the population has a great deal of confidence in the government has little meaning. How much is a "great deal"? We don't know whether this is higher or lower than the confidence held for other groups, such as big business and labor unions, or even whether confidence in government is higher or lower than it was last year or 10 years ago.

Analysis of individual variables is also problematic since relatively minor changes in the wording of questions may have sizable effects on the distribution of responses. In the GSS96 data file, some questions have alternate wordings, whereby half the sample (randomly selected) was asked one version of a question and half was asked the second version. This was a methodological study to determine how much changes in question wordings would affect the responses. Let's examine one example of this.

       Data File: **GSS96**
       ➤ *Task:* **Univariate**
➤ *Primary Variable:* **11) WELFARE $**
       ➤ *View:* **Pie**

This survey question asked for respondents' opinions on government spending on "welfare." Examine the percentage distributions for a moment. Then look at the alternate question, 17) WELFARE $2.

       Data File: **GSS96**
       Task: **Univariate**
➤ *Primary Variable:* **17) WELFARE $2**
       ➤ *View:* **Pie**

What may seem like slight changes in question wording had a huge effect on the percentage who selected each category. When asked about spending on welfare, only 15.6 percent thought the government was spending too little, but when asked about assistance to the poor, 56.2 percent thought too little was being spent.

Before seeing the data, almost any researcher would have accepted these questions as interchangeable indicators of the respondent's attitude toward helping the poor. After we see the data, we can, of course, speculate on how these questions must be tapping somewhat different attitudes, but such *ex post facto* analysis will not help us design indicators for another concept. If such minor changes in wording will make such a big change in the resulting distribution, how can social scientists ever develop adequate indicators of concepts?

Fortunately, as social scientists, we are primarily interested in the *relationship* between variables. It turns out that relationships between variables are usually much less sensitive than distributions of simple variables to changes in measurement techniques. Despite their limitations as descriptions, both of these question wordings are useful for comparing across groups. For example, we could use this variable to test the hypothesis that lower-income respondents favor spending to help the poor more than upper-income respondents do. First, let's use the cross-tabulation task and select WELFARE $ as the row variable and INCOME as the column variable.

Data File: **GSS96**
➤ *Task:* **Cross-tabulation**
➤ *Row Variable:* **11) WELFARE $**
➤ *Column Variable:* **76) INCOME**
➤ *View:* **Tables**
➤ *Display:* **Column %**

76) INCOME			
	Under 15K	15K-29,999	30K & over
11) Too little	23.2%	14.9%	11.1%
W Right	30.5%	28.7%	25.6%
E Too much	46.3%	56.4%	63.3%
L			
F			
A			
R			
E			
$			
TOTAL	100.0%	100.0%	100.0%

V=0.111**

Compare the percentages of people in different income groups who think too little is being spent. As the income levels increase, those who think too little is being spent on welfare decreases. Also look at the summary statistics and note that Cramer's V = .111 and the significance level = .000. Now make the alternate question using WELFARE $2 as the row variable and INCOME as the column variable.

Data File: **GSS96**
Task: **Cross-tabulation**
➤ *Row Variable:* **17) WELFARE $2**
➤ *Column Variable:* **76) INCOME**
➤ *View:* **Tables**
➤ *Display:* **Column %**

76) INCOME			
	Under 15K	15K-29,999	30K & over
17) Too little	66.5%	63.6%	50.6%
W Right	24.7%	22.5%	27.1%
E Too much	8.7%	13.8%	22.3%
L			
F			
A			
R			
E			
$			
2			
TOTAL	100.0%	100.0%	100.0%

V=0.121**

Compare the two groups again and then look at the summary statistics. For this relationship, Cramer's V = .121 and the significance level = .000. Note that, regardless of the wording difference, the *relationship* between income level and attitude toward spending to help the poor is the same. Our hypothesis would be supported with either variable: Lower-income respondents are more likely than upper-income respondents to think too little is being spent on helping the poor, and the results are statistically significant.

Studies that examine changes over time are also looking at relationships—time is one of the variables. For example, if we compare the distribution of attitudes toward spending on the poor in 1972 with the distribution of this attitude in 1996, we are looking at the effect of "time" on this attitude. Of course, for such a comparison to be valid, the same measure of attitude toward spending on the poor must be used each time.

Your turn.

Workbook exercises and software are copyrighted. Copying is prohibited by law.

1. After viewing the variable descriptions for the following variables, indicate the level of measurement by circling *nominal, ordinal,* or *interval/ratio*. Be sure to examine the categories used with each variable.

> ➤ *Data File:* **ANES96**
> ➤ *Task:* **Cross-tabulation**

17) SPENDING?	Nominal	Ordinal	Interval/ratio
35) WHO VOTE?	Nominal	Ordinal	Interval/ratio
49) POLITKNOW	Nominal	Ordinal	Interval/ratio
50) POLPART	Nominal	Ordinal	Interval/ratio
51) RELIGION	Nominal	Ordinal	Interval/ratio
54) AGE CATEGR	Nominal	Ordinal	Interval/ratio
56) FAMINCOME	Nominal	Ordinal	Interval/ratio

2. a. What is the description for 46) TRUST GOV?

_____

_____

_____

_____

_____

b. What concept do you think this variable is measuring?

c. What is the description for 45) CROOKED?

_____

_____

_____

_____

_____

d. Now let's cross-tabulate these two variables. Fill in the column percentages in the table that follows.

> Data File: **ANES96**
> Task: **Cross-tabulation**
> ➤ Row Variable: **46) TRUST GOV**
> ➤ Column Variable: **45) CROOKED?**
> ➤ View: **Tables**
> ➤ Display: **Column %**

	Quite Few	Not Many	Hardly Any
Most Time +	_____%	_____%	_____%
Only Some −	_____%	_____%	_____%

e. Let us assume for the moment that 46) TRUST GOV is a fairly valid measure of the concept you just specified. Based on this table, how valid is variable 45) CROOKED? in measuring this same concept? (Circle one.)

Very valid
Somewhat valid
Not very valid

Explain your answer.

3. Different kinds of research require different kinds of units of analysis (cases). For each of the following research questions, circle *yes* if it would be appropriate to use individuals as the units of analysis (cases) or *no* if it would not be appropriate.

a. Does federal spending on job training programs reduce the unemployment rate in the United States? (Circle one.)   Yes   No

b. Are students who do better in college more likely to find jobs when they graduate? (Circle one.)   Yes   No

c. Does economic development in nations lead to more democratic political institutions? (Circle one.)   Yes   No

d. Do people become more conservative on social issues as they grow older? (Circle one.)   Yes   No

e. Are women more supportive of sexual equality than men are? (Circle one.)   Yes   No

4. In the Canadian General Social Survey conducted by Statistics Canada, the following question asks respondents about church attendance:

Other than on special occasions, such as weddings, funerals or baptisms, how often did you attend services or meetings connected with your religion in the last 12 months? Was it . . .

> *At least once a week?*
> *At least once a month?*
> *A few times a year?*
> *At least once a year?*
> *Not at all*

a. Using the GSS96 data set, record the variable description and answer categories for 72) CH.ATTEND.

Description:

_____

Categories:

b. What concept is being measured by each of these two variables?

c. Which of these church attendance items is the better indicator? Why?

d. Record the description and answer categories of 99) CH.ATTEND!.

Description:

_____

Categories:

e. Let's compare the strengths and weaknesses of
this indicator with the one used by Statistics Canada.
Which one allows for the measurement of greater
variation? (Circle one.)   99) CH.ATTEND!

Statistics Canada question

f. Which one appears to be better at measuring the
frequency with which people *ordinarily* attend
religious services? (Circle one.)   99) CH.ATTEND!

Statistics Canada question

5. Find the percentages of survey respondents who would allow an abortion in
each of the situations specified in variables 40–45, and write these percentages in the blanks provided.

> ➤ *Data File:* **GSS96**
> ➤ *Task:* **Univariate**
> ➤ *Primary Variable:* **40) ABORT DEF (Repeat with 41) ABORT WANT,**
> **42) ABORT HLTH . . .)**
> ➤ *View:* **Pie**

40) ABORT DEF _____%

41) ABORT WANT _____%

42) ABORT HLTH _____%

43) ABORT NO$ _____%

44) ABORT RAPE _____%

45) ABORT SIGL _____%

Now, based on these results, discuss the following statement: "Most Americans approve of abortion. In a recent national random sample of adults, over 90% approved of abortion."

6. In addition to 11) WELFARE $ and 17) WELFARE $2 discussed in this exercise, alternate wordings of several other variables were included in the methodological study for GSS96. Let's examine some other examples of alternate wordings.

    a. Provide the variable description and distribution for the following variable:

    > Data File: **GSS96**
    > Task: **Univariate**
    > ➤ Primary Variable: **6) ENVIRON. $**
    > ➤ View: **Pie**

    Description:

    _____

    _____

    In the following table, fill in the appropriate frequency and percentage distributions.

	Frequency	Percent
Too little	_____	_____%
Right	_____	_____%
Too much	_____	_____%

b. Provide the variable description and distribution for the following variable.

> Data File: **GSS96**
> Task: **Univariate**
> ➤ Primary Variable: **12) ENVIRON.$2**
> ➤ View: **Pie**

Description:

_____

Fill in the following table.

	Frequency	Percent
Too little	_____	_____%
Right	_____	_____%
Too much	_____	_____%

c. Compare the results for these two variables. Do you think it is legitimate to consider these two variables as indicators of the same concept? (Circle one.)   Yes   No

d. If you answered YES, provide a definition of the concept. If you answered NO, provide definitions of the two concepts being measured.

e. Now examine the relationship between 75) POL. VIEW and the first of these environmental variables. The hypothesis is: Liberals are more likely to think too little is being spent on the environment than are conservatives.

> Data File: **GSS96**
> ➤ Task: **Cross-tabulation**
> ➤ Row Variable: **6) ENVIRON. $**
> ➤ Column Variable: **75) POL. VIEW**
> ➤ View: **Tables**
> ➤ Display: **Column %**

Fill in the column percentages in the following table, and then fill in the Cramer's V and the significance level for the relationship.

	Liberal	Moderate	Conservative
Too little	_____%	_____%	_____%
Right	_____%	_____%	_____%
Too much	_____%	_____%	_____%

V = _____

Prob. = _____

Is the hypothesis supported or rejected?
(Circle one.)   Supported   Rejected

f. Now examine the relationship between 75) POL. VIEW and the second of these environmental variables. The hypothesis is still: Liberals are more likely to think too little is being spent on the environment than are conservatives.

> *Data File:* **GSS96**
> *Task:* **Cross-tabulation**
> ➤ *Row Variable:* **12) ENVIRON.$2**
> ➤ *Column Variable:* **75) POL. VIEW**
> ➤ *View:* **Tables**
> ➤ *Display:* **Column %**

Fill in the column percentages in the following table, and then fill in the Cramer's V and the significance level for the relationship.

	Liberal	Moderate	Conservative
Too little	_____%	_____%	_____%
Right	_____%	_____%	_____%
Too much	_____%	_____%	_____%

V = _____

Prob. = _____

Is the hypothesis supported or rejected?
(Circle one.)   Supported   Rejected

g. In testing this hypothesis (liberals are much more likely too think too little is being spent on the environment than conservatives are), did it make a difference in the results which indicator of environmental spending attitudes (ENVIRON.$ or ENVIRON.$2) you selected? (Circle one.)   Yes   No

Explain your answer below.

7. Another set of variables with alternative wording in the GSS96 file is 9) CRIME $ and 15) CRIME $2. Let's examine each of these variables.

a. Provide the description and distribution for the following variable:

> Data File: **GSS96**
> ➤ Task: **Univariate**
> ➤ Primary Variable: **9) CRIME $**
> ➤ View: **Pie**

Description: 9) CRIME $

_____

_____

Distribution:

	Frequency	Percent
Too little	_____	_____%
Right	_____	_____%
Too much	_____	_____%

b. Provide the description and distribution for the following variable:

> Data File: **GSS96**
> Task: **Univariate**
> ➤ Primary Variable: **15) CRIME $2**
> ➤ View: **Pie**

Description: 15) CRIME $2

_____

_____

Distribution:

	Frequency	Percent
Too little	_____	_____%
Right	_____	_____%
Too much	_____	_____%

c. Compare the results for these two variables. Do you
   think it is legitimate to consider these two variables as
   indicators of the same concept? (Circle one.)   Yes   No

d. If you answered YES, provide a definition of the concept. If you answered
   NO, provide definitions of the two concepts being measured.

8. Now test the hypothesis that there will be a relationship between 31) FEAR
   WALK and the first of these crime spending variables. The hypothesis is:
   Those who are afraid to walk alone at night are more likely to think too little
   is being spent on crime spending than are individuals who are not afraid.

Data File:	**GSS96**
➤ Task:	**Cross-tabulation**
➤ Row Variable:	**9) CRIME $**
➤ Column Variable:	**31) FEAR WALK**
➤ View:	**Tables**
➤ Display:	**Column %**

a. Fill in the column percentages in the following table, and then fill in the
   Cramer's V and the significance level for the relationship.

	Yes	No
Too little	_____%	_____%
Right	_____%	_____%
Too much	_____%	_____%

$$V = \underline{\hspace{2cm}}$$

$$Prob. = \underline{\hspace{2cm}}$$

Is the hypothesis supported or rejected?
(Circle one.)   Supported   Rejected

Now test the hypothesis that there will be a relationship between 31) FEAR WALK and the second of these crime spending variables. The hypothesis is: Those who are afraid to walk alone at night are more likely to think too little is being spent on crime spending than are individuals who are not afraid.

> Data File: **GSS96**
> Task: **Cross-tabulation**
> ➤ Row Variable: **15) CRIME $2**
> ➤ Column Variable: **31) FEAR WALK**
> ➤ View: **Tables**
> ➤ Display: **Column %**

b. Fill in the column percentages in the following table, and then fill in the Cramer's V and the significance level for the relationship.

	Yes	No
Too little	_____%	_____%
Right	_____%	_____%
Too much	_____%	_____%

$$V = \underline{\hspace{2cm}}$$

$$Prob. = \underline{\hspace{2cm}}$$

Is the hypothesis supported or rejected?
(Circle one.)   Supported   Rejected

c. In testing this hypothesis (those who are afraid to walk alone at night are more likely to think that too little is being spent on crime spending), did it make a difference in the results which indicator of crime spending attitudes (CRIME $ or CRIME $2) you selected? (Circle one.)   Yes   No

Explain your answer below.

9. ABORT INDX is a support for abortion index consisting of the number of YES answers respondents gave to six abortion questions (40–45). Let's examine Cronbach's alpha reliability coefficient for this index. Remember that we want alpha to be at least .7 in order to decide that the variables constituting an index provide a reliable measure.

> Data File: **GSS96**
> ➤ Task: **Correlation**
> ➤ Select Variables: **40) ABORT DEF**
> **41) ABORT WANT**
> **42) ABORT HLTH**
> **43) ABORT NO$**
> **44) ABORT RAPE**
> **45) ABORT SIGL**

a. What is the alpha for this set of variables?   Alpha = _____

b. On the basis of the value of alpha, would you
   conclude that the support for abortion index
   (ABORT INDX) is reliable or unreliable?
   (Circle one.)   Reliable   Unreliable

In the area of abortion attitudes, we might want to distinguish two clusters of attitudes: one set that broadly concerns emergencies and another set that is concerned with less threatening matters. What happens to the degree of reliability if we split these six variables into two sets of three? Let's consider the three abortion questions that concern emergencies first.

> Data File: **GSS96**
> Task: **Correlation**
> ➤ Select Variables: **40) ABORT DEF**
> **42) ABORT HLTH**
> **44) ABORT RAPE**

c. What is the alpha for this set of variables?   Alpha = _____

d. On the basis of the value of alpha, would you
   conclude that this support for abortion index
   is reliable or unreliable? (Circle one.)   Reliable   Unreliable

Now let's look at the correlations for the other three abortion questions.

> Data File: **GSS96**
> Task: **Correlation**
> ➤ Select Variables: **41) ABORT WANT**
> **43) ABORT NO$**
> **45) ABORT SIGL**

a. What is the alpha for this set of variables?   Alpha = _____

b. On the basis of the value of alpha, would you
   conclude that this support for abortion index
   is reliable or unreliable? (Circle one.)   Reliable   Unreliable

4

Selecting Cases

OVERVIEW

In this exercise, you will learn more about problems that can cause bias in samples, and you will also see that some sampling problems probably do not seriously distort the accuracy of the results.

BEFORE YOU BEGIN

Please make sure you have read Chapter 4 in the textbook and can answer the following review questions (you need not write any answers):

1. What is the difference between a census and a sample?

2. What is the difference between a parameter and a statistic?

3. If the confidence interval for a sample is plus or minus 3 percentage points and the confidence level is 95 percent, what does this mean in terms of the results?

4. Suppose you have a list of all students in your college and you want to take a sample of them. How would you go about selecting (a) a simple random sample and (b) a systematic random sample?

5. Describe the general process by which you would select a probability sample of people in a city for a telephone survey.

6. Describe how each of the following sources of bias can distort random samples: nonresponse bias, selective availability bias, areal bias.

7. What is a SLOPS, and how trustworthy are the results of such surveys?

The first step in selecting cases is to define the relevant population. When one is primarily interested in *describing* a particular group of units, determining the population is straightforward. A magazine, for example, might be interested in the opinions of its current subscribers—the appropriate population is individuals who subscribe to the magazine. Or a college might want to determine how its students budget their time—students enrolled at the school would be the population of interest. A city might want to evaluate the utilization of recreational facilities—the available facilities might be used as the cases.

When research shifts to *testing ideas* rather than describing groups, defining the appropriate population is not so easy. For example, a researcher testing a theory about the effect of organizational structure on decision making would probably want to define the relevant population as all organizations over all time. Similarly, someone interested in the effect of new and contradictory information on attitude change might want to generalize the results to all humans over all time. Clearly, however, there is no way to take a census of either of these populations or even to draw scientific samples. Consequently, researchers settle for working with a subset of the population of interest.

For example, the organization researcher might define the population as all business firms within a particular geographic area at the current time. The attitude change researcher might define the population as individuals over age 18 currently residing in the United States. Other researchers testing these same theories might choose quite different populations. In fact, one type of replication research is to test the same hypotheses in a different population. Thus, a researcher might replicate the organizational study by using national charitable organizations as the relevant population.

Having defined the appropriate population, the next step is to determine how cases are to be selected from this population. For many research questions, using a census of the units is more appropriate than selecting a sample. If an instructor were interested in the opinions and attitudes of students in a particular class, he or she should simply take a census of the class—that is, include all members in the study. There is no reason to select a sample, since the entire population is accessible and relatively small—why introduce unnecessary sources of inaccuracy? If one is studying state governments, governments of all 50 states would be included, unless the data collection process was quite complicated. In short, samples are used only when taking a census would be more difficult.

If a sample is to be used, researchers must determine how to select the sample. If we wish to generalize the results from the sample to the population, then we must use a probability (or random) sample. With such samples, we can determine the exact probability that a particular case in the population is included in the sample. Using probability theory, we can then generalize from the observed sample to the population. This process of inferring from a sample to a population will be covered in detail when you take a statistics course. For now, you can simply assume that, when a proper sample is selected, the sample statistic provides the best estimate of the population parameter. For example, if the proportion of females in the sample is .63 (sample statistic), then our best estimate of the proportion of females in the population is .63 (population parameter).

An important property of probability samples is that two or more random samples of the same population can be combined and this combined sample is also a probability sample. This can be very useful. Suppose, for example, you find that a substantial number of cases in your sample are no longer members of the selected population—perhaps they have moved to an area not included in your sampling frame. As a result, your sample will have many fewer cases than you had planned. You can replace these cases by simply drawing another sample, including sufficient cases to complete the original sample, and combine these data with your first sample (you would, of course, want to drop any duplicates that appear).

The problem with an *improperly* selected sample is that estimates of the population parameters are almost certain to be incorrect. Let's look at an example. Suppose you administered a survey to those attending Sunday services at several churches. Let's take a look at church attendance.

These results indicate that unless some unusual Sunday is selected, we would expect less than half the population to be in church. But the real problem is that those who are in church will differ in many ways from those not in church. Let's look at some of these differences. First, let's find out what the percentages of males and females are in the total sample.

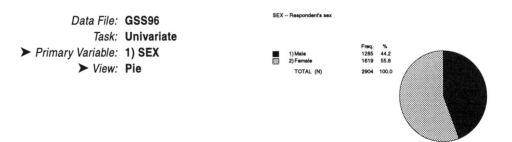

The percentage of males in the total sample is 44.2. Now let's look at the sex distribution of church attendees.

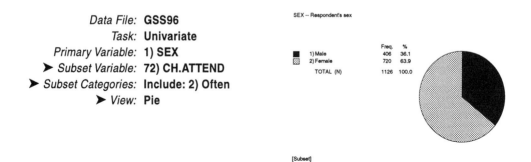

		Data File:	**GSS96**
		Task:	**Univariate**
		Primary Variable:	**1) SEX**
➤		Subset Variable:	**72) CH.ATTEND**
➤		Subset Categories:	**Include: 2) Often**
	➤	View:	**Pie**

The option for selecting a subset variable is located on the same screen you use to select other variables. For this example, select 72) CH.ATTEND as a subset variable. A window will appear that shows you the categories of the subset variable. Mark the box 2) OFTEN as your subset category and then choose the "include" option. Click [OK] and continue as usual. (Note: Subset variables remain selected until you manually delete them or until you return to the main menu.)

The resulting pie chart contains information on only those who were coded 2 on 72) CH.ATTEND—those who attend church often. Among those who attend church frequently, only 36.1 percent are males. So we see one way in which this sample is quite biased: Males are likely to be underrepresented. We can think of several other ways in which such a sample would probably be biased. We might expect younger individuals who are single to be underrepresented. Or perhaps individuals from different regions will be differentially represented.

Even if you selected a proper probability sample of churchgoers, you could not generalize your results to the general population. In fact, if you simply collected data at some churches or collected data on a particular Sunday, you would not have a proper sample of churchgoers and you could not even generalize your results to churchgoers.

Another form of sampling bias can result from the type of interviewing process that is selected. Some surveys, such as the U.S. General Social Survey, use face-to-face interviews, whereas other surveys use telephone interviews. Let's see if there is any serious bias generated by using telephone interviews. First, using the UNIVARIATE task, determine what percentage of the respondents have telephones.

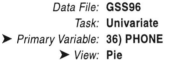

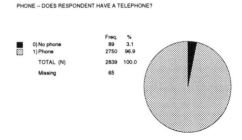

PHONE -- DOES RESPONDENT HAVE A TELEPHONE?

		Freq.	%
■	0) No phone	89	3.1
▨	1) Phone	2750	96.9
	TOTAL (N)	2839	100.0
	Missing	65	

Data File: **GSS96**
Task: **Univariate**
➤ *Primary Variable:* **36) PHONE**
➤ *View:* **Pie**

If you are continuing from the previous example, make sure that you have deleted the subset variable that was used in the prior example. Again, subset variables remain selected until you manually delete them or until you return to the main menu. Remember, the [Clear All] button deletes all variable selections.

We can see that 89 cases—only 3.1 percent of the sample—did not have a phone. If NORC had conducted phone rather than personal interviews, the interviewers would not have been able to reach these individuals who didn't have phones. While this small percentage of cases would not make much difference in the results, let's nevertheless investigate who we are likely to undersample if we select only cases who have telephones. Perhaps those with low incomes would be undersampled, since they are probably somewhat less likely to have telephones. Let's first look at the distribution of income in the total sample.

Data File: **GSS96**
Task: **Univariate**
➤ *Primary Variable:* **77) R.INCOME**
➤ *View:* **Pie**

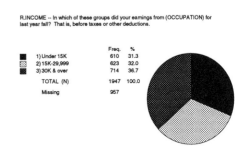

R.INCOME -- In which of these groups did your earnings from (OCCUPATION) for last year fall? That is, before taxes or other deductions.

		Freq.	%
■	1) Under 15K	610	31.3
▨	2) 15K-29,999	623	32.0
▨	3) 30K & over	714	36.7
	TOTAL (N)	1947	100.0
	Missing	957	

As we can see, each category has about one-third of the sample. Now let's see what will happen if we look at the income distribution of only those individuals who have telephones.

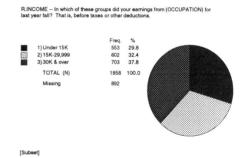

Data File: **GSS96**
Task: **Univariate**
Primary Variable: **77) R.INCOME**
➤ Subset Variable: **36) PHONE**
➤ Subset Categories: **Include: 1) Phone**
➤ View: **Pie**

Only cases with phones will be included in the distribution. We can see that excluding cases with no telephone makes only a small difference in the income distribution. For example, the percentage in the lowest income category changes from 31.3 to 29.8, or less than two percentage points.

Similarly we might expect older individuals to be undersampled in a phone survey. Let's look at the distribution of age in the total sample.

Data File: **GSS96**
Task: **Univariate**
➤ Primary Variable: **67) OVER 50**
➤ View: **Pie**

Again, make sure you delete the subset variable that was selected in the previous example.

Let's now look at the distribution of age among respondents with telephones.

Data File: **GSS96**
Task: **Univariate**
Primary Variable: **67) OVER 50**
➤ Subset Variable: **36) PHONE**
➤ Subset Categories: **Include: 1) Phone**
➤ View: **Pie**

When individuals with phones are excluded, the change is less than 1 percentage point in either of these categories. This suggests that conducting interviews over the telephone rather than using personal interviews would not have a serious effect on the results, assuming that a proper probability sample were selected (and that the nonresponse rate did not substantially increase).

Bias also can be introduced when cases selected for the sample refuse to participate. For example, in the 1996 General Social Survey, 910 individuals in the selected sample failed to complete the interview for one reason or another. To the extent that these individuals differ from those who completed the interview, the sample will be biased. Or, if a sizable number of individuals refuses to answer a particular question, any analysis involving that question may fail to reflect the population parameters. Occasionally, individuals are dropped because their answers are clearly frivolous. Sometimes, interviews are not completed because respondents become too hostile or are obviously incapable of understanding the questions. In telephone interviews, a respondent may simply hang up in the middle of the interview.

In order to help identify cases who may be providing incorrect answers, interviewers are frequently asked to provide information about the general ambience of the interview. For example, the interviewers for the GSS were asked to rate each respondent's level of comprehension. Let's look at the results.

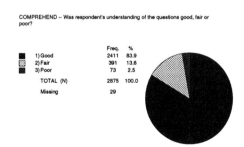

Data File: **GSS96**
Task: **Univariate**
➤ *Primary Variable:* **32) COMPREHEND**
➤ *View:* **Pie**

COMPREHEND -- Was respondent's understanding of the questions good, fair or poor?

		Freq.	%
■	1) Good	2411	83.9
▨	2) Fair	391	13.6
▨	3) Poor	73	2.5
	TOTAL (N)	2875	100.0
	Missing	29	

Only a small percent (2.5%) of respondents had difficulty understanding the questions. The likely effect of retaining individuals with poor understanding is introducing random "noise" into the survey—their answers are more likely to contain a random component than are answers of other respondents.

In addition, interviewers were asked to report whether respondents were cooperative or hostile. Let's look at the results.

Data File: **GSS96**
Task: **Univariate**
➤ *Primary Variable:* **33) ATTITUDE?**
➤ *View:* **Pie**

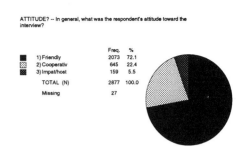

The percent who were impatient or hostile is also very small.

Even a properly selected sample will not be perfectly accurate in estimating population parameters. Generalizing from samples to population is a probabilistic, not a deterministic, process. For example, based on the GSS sample, our best estimate of the percent of males in the population is 44.2. Even if the response rate had been perfect, the true population parameter would likely be somewhat different from this estimate. If we select another sample of 2,904 cases using similar procedures, the percent of males is likely to be slightly different. If we select yet another sample, we will get yet another estimate. In a statistics course, you will learn how to evaluate the accuracy of such estimates.

Your turn.

NAME:

COURSE:

DATE:

EXERCISE

4

Workbook exercises and software are copyrighted. Copying is prohibited by law.

WORKSHEET

1. The difference between population values for a variable and the sample values for the same variable is *sampling error*. For example, if the average age in the population of a state is 45.0 and the average age from a sample of this population is 46.5, then the sampling error is 1.5.

    To demonstrate this, we combined all the cases from the 1990 to 1994 General Social Surveys—a total of 7,487 people—to serve as a "population." Then we took a random sample of 750 of these respondents using random digits generated through the full version of MicroCase. We will use two variables (education and age) to demonstrate.

    Below are the averages for education and age for the population (the combined samples) and the sample. For each situation, specify the sampling error.

	Population	Sample	Sampling Error
Education	13.02	13.07	_____
Age	45.91	45.46	_____

    Would you say that the sample did a good job of representing accurately the population for these two variables? (Circle one.)   Yes   No

2. Suppose that you were doing a survey and that, instead of selecting a random sample and using telephone or personal interviews, you decided to find your respondents at the local tavern. Using the GSS96 data, we can see how those who go to bars twice a month or more often differ from those who don't go to bars.

    > *Data File:* **GSS96**
    > *Task:* **Univariate**
    > *Primary Variable:* **55) SOC. BAR**
    > *View:* **Pie**

a. Fill in the following table.

	Frequency	Percent
2+/month	_____	_____
1-/month	_____	_____

b. Perhaps females would be undersampled if we selected the sample from bars. Let's first check the percentages of males and females and fill in the table that follows.

> Data File: **GSS96**
> Task: **Univariate**
> ➤ Primary Variable: **1) SEX**
> ➤ View: **Pie**

	Frequency	Percent
Male	_____	_____
Female	_____	_____

In the total sample, what is the percentage of females?   _____%

c. Let's compare the distribution of sex in the total sample with the subset who frequent bars. Fill in the table that follows.

> Data File: **GSS96**
> Task: **Univariate**
> Primary Variable: **1) SEX**
> ➤ Subset Variable: **55) SOC. BAR**
> ➤ Subset Categories: **Include: 1) 2+/month**
> ➤ View: **Pie**

Subset with value 2+/month

	Frequency	Percent
Male	_____	_____
Female	_____	_____

Among those who go to bars frequently, what percentage is female?   _____%

What does this suggest about the distribution of sex among a sample drawn from local taverns? (Support your answer with results from the preceding analyses.)

d. People who frequent bars probably differ in many other ways from the general population. Select a variable of your choice (ideally a variable that describes a "demographic" characteristic, rather than an "attitudinal" variable) that you think will be different for people who go to bars. Then conduct a similar analysis to that done above.

Name of the variable: _____

Description of the variable:

_____

_____

e. How do you expect this variable to be different in terms of people who go to bars and the general population?

> Data File: **GSS96**
> Task: **Univariate**
> ➤ Primary Variable: **(the variable you selected above)**
> ➤ View: **Pie**

Provide the distribution of the variable in the *total* sample —use only as many rows as necessary.

Category Name	Frequency	Percent
_____	_____	_____
_____	_____	_____
_____	_____	_____
_____	_____	_____
_____	_____	_____
_____	_____	_____

f. Obtain the distribution of the variable in the subset that goes to bars frequently. In addition to filling in the results below, attach a printout of your analysis. (If your computer is not connected to a printer or if you have been instructed not to use the printer, just skip these printing instructions.)

> Data File: **GSS96**
> Task: **Univariate**
> Primary Variable: **(the variable you selected)**
> ➤ Subset Variable: **55) SOC. BAR**
> ➤ Subset Categories: **Include: 1) 2+/month**
> ➤ View: **Pie**

Subset with value 2+/month

Category Name	Frequency	Percent
_____	_____	_____
_____	_____	_____
_____	_____	_____
_____	_____	_____
_____	_____	_____
_____	_____	_____

g. Discuss the implications of these results for obtaining your respondents from taverns. (Support your answer with the results of your analysis.)

3. Let's see if the distribution of educational attainment would be affected if only individuals with telephones were included in the sample.

> Data File: **GSS96**
> Task: **Univariate**
> ➤ Primary Variable: **64) DEGREE**
> ➤ View: **Pie**

a. Fill in the following table for this distribution.

	Frequency	Percent
Not hi sch		
High sch		
Some coll		

b. Now repeat this analysis using 36) PHONE as the subset variable and fill in the table that follows.

> Data File: **GSS96**
> Task: **Univariate**
> Primary Variable: **64) DEGREE**
> ➤ Subset Variable: **36) PHONE**
> ➤ Subset Categories: **Include: 1) Phone**
> ➤ View: **Pie**

Subset with value phone

	Frequency	Percent
Not hi sch		
High sch		
Some coll		

c. Describe how using the phone interviews rather than personal interviews would affect the distribution of education in the observed sample. (Provide support for your answer.)

4. Now let's investigate how respondents' education level is related to compre-
    hension of questions. This time we will use the CROSS-TABULATION task.

                    Data File: **GSS96**
                  ➤ Task: **Cross-tabulation**
            ➤ Row Variable: **64) DEGREE**
         ➤ Column Variable: **32) COMPREHEND**
                    ➤ View: **Tables**
                 ➤ Display: **Column %**

a. Fill in the column percentages in the following table.

	Good	Fair	Poor
Not hi school	_____%	_____%	_____%
High sch	_____%	_____%	_____%
Some coll	_____%	_____%	_____%

b. Similarly, let's investigate how respondents' education level is related to
    attitude toward the interview.

                    Data File: **GSS96**
                       Task: **Cross-tabulation**
               Row Variable: **64) DEGREE**
         ➤ Column Variable: **33) ATTITUDE?**
                    ➤ View: **Tables**
                 ➤ Display: **Column %**

Fill in the column percentages in the following table.

	Friendly	Cooperative	Impat/host
Not hi school	_____%	_____%	_____%
High sch	_____%	_____%	_____%
Some coll	_____%	_____%	_____%

c. Assuming that individuals who are hostile or impatient or who fail to understand the questions are most likely to break off an interview, how would the distribution of educational attainment in the sample be affected?

5. On presidential election day, newscasters report results of exit polls based on surveys of voters leaving the polling area. In addition to reporting results on votes, the newscasters might also report the views of voters on various political issues and they might discuss the demographic characteristics of voters. These exit poll results are appropriate for describing the relevant population (voters), but the results might not be as accurate in reflecting the views of the general public—both voters and non-voters. Non-voters might differ from voters in terms of political attitudes and demographic character-istics.

Let's use the ANES96 data file in order to investigate this. (Note: Many who claimed that they voted did not actually vote. Although this does cause a problem for the analysis we're doing here, we will ignore this issue for the present.)

a. Let's first obtain the distribution of attitudes for 18) MED. COST.

> *Data File:* **ANES96**
> *Task:* **Univariate**
> *Primary Variable:* **18) MED. COST**
> *View:* **Pie**

Fill in the information in the following table.

	Frequency	Percent
1) Gov't pay	_____	_____
2) Middle	_____	_____
3) Priv pay	_____	_____

b. Now let's look at the distribution of attitudes on this question for just those who said they voted.

> Data File: **ANES96**
> Task: **Univariate**
> Primary Variable: **18) MED. COST**
> ➤ Subset Variable: **34) VOTED?**
> ➤ Subset Categories: **Include: 1) Yes**
> ➤ View: **Pie**

Fill in the information in the following table.

Subset with value YES

	Frequency	Percent
1) Gov't pay	_____	_____
2) Middle	_____	_____
3) Priv pay	_____	_____

c. Compare the distribution for the total sample with the distribution for those who said they voted. If we used a sample of voters to represent the views of the general public, what effect would it have on our impression of public attitudes concerning this issue? First, would we overestimate or underestimate the level of support for governmental assistance for medical care? Second, do you think this difference is *substantively* significant? That is, do you think the percentage difference between those who voted and those who did not vote is large enough to be concerned with? (Support your answer using the results of your analysis.)

d. Would using the views of voters to represent public opinion cause a problem in examining views on the issue of prayer in public schools? First, obtain the distribution for the total sample.

> Data File: **ANES96**
> Task: **Univariate**
> ➤ Primary Variable: **40) SC. PRAYER**
> ➤ View: **Pie**

Fill in the information in the following table.

	Frequency	Percent
No prayer	_____	_____
Silent prayer	_____	_____
Prayer time	_____	_____
Required prayer	_____	_____

e. Now let's look at the distribution of attitudes on this question for just those who said they voted.

> Data File: **ANES96**
> Task: **Univariate**
> Primary Variable: **40) SC. PRAYER**
> ➤ Subset Variable: **34) VOTED?**
> ➤ Subset Categories: **Include: 1) Yes**
> ➤ View: **Pie**

Fill in the information in the following table.

Subset with value YES

	Frequency	Percent
No prayer	_____	_____
Silent prayer	_____	_____
Prayer time	_____	_____
Required prayer	_____	_____

f. Compare the distribution for the total sample with the distribution for those who said they voted. If we used a sample of voters to represent the views of the general public, what effect would it have on our impression of public attitudes concerning this issue? Would we overestimate or underestimate the level of support for prayer in public schools? (Use the results of the analysis to support your answer.)

6. A researcher wants to generalize from his survey results to the population that received his questionnaire. When asked if the returned questionnaires could be considered a random sample of the population, he replies, "We sent out 100,000 questionnaires and got only 5,000 back. What could be more random than that?" Is he right? Why or why not?

5a

Control Variables

OVERVIEW

In Exercise 5a, you will learn how to examine a relationship between two variables while controlling for the effects of another variable—and you will see why this is important when we try to disentangle relationships among variables. Exercise 5b is an optional assignment (ask your instructor) that looks at more advanced issues on this topic. In this exercise, you will gain experience in using regression to examine the simultaneous effects of two or more independent variables on a dependent variable, and you will see how this is useful in assessing models of what causes the variance in the dependent variable.

BEFORE YOU BEGIN

Please make sure you have read Chapter 5 in the textbook and can answer the following review questions (you need not write any answers):

1. Describe the three criteria of causation (time order, etc.).

2. How are independent, dependent, intervening, and antecedent variables related to each other?

3. What does multiple causation mean?

4. What are causal models, and how is regression used in assessing causal models?

5. If the beta coefficient for an independent variable in regression is .50, what does this mean? (Assume that it is statistically significant.)

6. If a squared multiple regression correlation (R^2) is .80, what does this mean?

7. How can regression be useful in identifying spurious relationships?

8. How can regression help in sorting out the effects of suppressor variables?

9. What are dummy variables, and how are they used in regression?

Before we look at control variables in cross-tabulation, let's take a closer look at the variables we'll be using.

➤ *Data File:* **GSS96**
➤ *Task:* **Cross-tabulation**
➤ *Row Variable:* **96) R.INCOME!**
➤ *Column Variable:* **77) R.INCOME**
➤ *View:* **Tables**
➤ *Display:* **Frequency**

77) R.INCOME

		Under 15K	15K-29,999	30K & over
96)	Under 1K	28	0	0
R	1K-2999	88	0	0
i	3K-3999	39	0	0
N	4K-4999	36	0	0
C	5K-5999	39	0	0
O	6K-6999	41	0	0
M	7K-7999	36	0	0
E	8K-9999	73	0	0
!	10K-12499	118	0	0
	12.5-14999	112	0	0
	TOTAL	610	623	714

V=1.000**

> Although you are generally instructed to display column percentaging in your cross-tabula-
> tion analysis, note that here we are looking at the frequency distribution.

Variable 77) R.INCOME is based on exactly the same question as variable 96) R. INCOME!. The only difference is that variable 96 includes all answer categories made available to respondents for reporting their own annual income, whereas variable 77 was created by combining a number of these categories. The table shows you which categories of 96) R.INCOME! have been combined to form the categories for 77) R.INCOME. For example, categories from UNDER 1K to 12.5–14999 on variable 96) R.INCOME! were placed in the category UNDER 15K on variable 77) R.INCOME.

The procedure of combining response categories for a variable is called *col-lapsing*—77) R.INCOME is a *collapsed* version of 96) R.INCOME!. In this data set, all variables with an exclamation point are uncollapsed versions of variables (many of these also are included in a collapsed form). The uncollapsed versions are located toward the end of the variable list. It may have crossed your mind that collapsing seems to go against the idea of trying to have as much variation as possible in our variables since variable 77 has only three categories while variable 96 has 21. Why do we do this? Obtain the following cross-tabulation table and look at the column percentages.

Data File: **GSS96**
Task: **Cross-tabulation**
Row Variable: **96) R.INCOME!**
➤ *Column Variable:* **92) EDUCATION!**
➤ *View:* **Tables**
➤ *Display:* **Column %**

92) EDUCATION!

		No school	3rd grade	4th grade	5th grade	6th grade
96)	Under 1K	0.0%	-	-	0.0%	0.0%
R	1K-2999	0.0%	-	-	0.0%	0.0%
i	3K-3999	0.0%	-	-	0.0%	0.0%
N	4K-4999	0.0%	-	-	0.0%	0.0%
C	5K-5999	0.0%	-	-	0.0%	16.7%
O	6K-6999	0.0%	-	-	50.0%	0.0%
M	7K-7999	0.0%	-	-	0.0%	0.0%
E	8K-9999	0.0%	-	-	0.0%	16.7%
!	10K-12499	100.0%	-	-	0.0%	33.3%
	12.5-14999	0.0%	-	-	0.0%	0.0%
	TOTAL	100.0%	-	-	100.0%	100.0%

V=0.153**

You need to use the scroll bar to move around in this huge table. Even a very experienced professional analyst would not be able to make much sense of this table. Many of the columns have very few cases, so the percentages for the most part are meaningless. This is why we collapse variables when using tabular analysis—to get enough cases in the categories to be able to interpret the results.

Because we will be using cross-tabulation in this exercise, virtually all of the variables are in collapsed form. While collapsing will change the *distribution* of the variable, such changes rarely affect *relationships* with other variables as we saw in Exercise 2b. (In Chapter 7 of the text, you'll learn the principles of collapsing variables.)

Now let's look at an example of a spurious relationship. We will begin by cross-tabulating sexual frequency by marital status and filling in the column percentages in the table that follows.

	Data File:	**GSS96**
	Task:	**Cross-tabulation**
➤	Row Variable:	**57) SEX FREQ**
➤	Column Variable:	**35) MARITAL**
➤	View:	**Tables**
➤	Display:	**Column %**

35) MARITAL	Married	Div/widowd	Nev.marry
57) -Monthly	9.9%	49.2%	32.1%
Monthly+	90.1%	50.8%	67.9%
TOTAL	100.0%	100.0%	100.0%

V=0.385**

The row variable tells us how often a respondent has had sex during the previous 12 months. In this form, the variable has been collapsed into those who had sex less often than once a month and those who had sex at least once a month or more often. It should surprise no one that married people were more apt to have sex at least once a month than were people in the other two categories. However, it does seem unusual that the difference between the never married and the widowed and divorced group should be so great (around a 17 percent difference). Let's focus on just those two groups.

	Data File:	**GSS96**
	Task:	**Cross-tabulation**
	Row Variable:	**57) SEX FREQ**
	Column Variable:	**35) MARITAL**
➤	Subset Variable:	**35) MARITAL**
➤	Subset Categories:	**Exclude: 1) Married**
➤	View:	**Tables**
➤	Display:	**Column %**

35) MARITAL	Div/widowd	Nev.marry
57) -Monthly	49.2%	32.1%
Monthly+	50.8%	67.9%
TOTAL	100.0%	100.0%

V=0.173**

The option for selecting a subset variable is located on the same screen you use to select other variables. For this example, select 35) MARITAL as a subset variable. A window will appear that shows you the categories of the subset variable. Mark the box 1) MARRIED and then choose the "exclude" option. (This means that all other categories for this variable will be included in the analysis.) Click [OK] and continue on as usual. Remember, subset variables remain selected until you manually delete them or until you return to the main menu.

We can now focus on the difference between the divorced/widowed and never-married groups because our analysis excludes people who are married. This table shows a percentage difference of 17.1 in the first row, and the N's (number of cases) reported at the bottom are large enough that we can have some confidence in the stability of the percentages. You can see that Cramer's V is .173 and is highly significant.

We have now observed a relationship between marital status and frequency of sex among those who are not married. But could this relationship be spurious? Is there some other variable that could be creating this difference? If so, this other variable must precede these two variables in time. Perhaps age might be this variable. Age has an effect on both of these variables—the never-married group is probably younger and younger individuals are probably more likely to have sex.

Let's diagram this argument:

AMONG THE NOT MARRIED

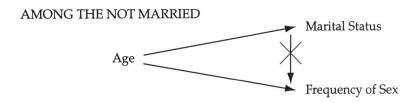

The position of the variables from left to right indicates time order. The arrows indicate causal relationships. The arrow at the right with the X across it represents a relationship that disappears when we control for age. Now let's test this argument by controlling for age.

Data File: **GSS96**
Task: **Cross-tabulation**
Row Variable: **57) SEX FREQ**
Column Variable: **35) MARITAL**
Subset Variable: **35) MARITAL**
Subset Categories: **Exclude: 1) Married**
➤ Control Variable: **67) OVER 50**
➤ View: **Tables (Under 50)**
➤ Display: **Column %**

Control: Under 50

		35) MARITAL	
		Div/widowd	Nev.marry
57)	-Monthly	27.1%	28.6%
SEX	Monthly+	72.9%	71.4%
FREQ			
	TOTAL	100.0%	100.0%

V=0.017

If you are continuing from the previous example, and you haven't returned to the main menu, simply add a control variable to your previous variable selections. The option for selecting a control variable is located on the same screen you use to select other variables. For this example, select 67) OVER 50 as a control variable and then click [OK] to continue. Separate tables for each of the OVER 50 categories will now be shown for the SEX FREQ and MARITAL cross-tabulation.

The first table will show the relationship between marital status and frequency of sex for those unmarried respondents under 50. As we can see, there is virtually no difference between the divorced/widowed and never-married groups among those under 50. You can see that Cramer's V is .017 and clearly not significant.

Let's move on to the second control table, which limits the analysis to those who are 50 and over.

Data File:	**GSS96**
Task:	**Cross-tabulation**
Row Variable:	**57) SEX FREQ**
Column Variable:	**35) MARITAL**
Subset Variable:	**35) MARITAL**
Subset Categories:	**Exclude: 1) Married**
Control Variable:	**67) OVER 50**
➤ View:	**Tables (50 & Over)**
➤ Display:	**Column %**

Control: 50 & Over

		35) MARITAL	
		Div/widowd	Nev.marry
57) SEX FREQ	-Monthly	72.8%	70.0%
	Monthly+	27.2%	30.0%
TOTAL		100.0%	100.0%

V=0.021

If you are continuing from the previous example, click the appropriate button at the bottom of the task bar to look at the second (or "next") partial table for those who are 50 and over.

As can be seen from this table, among those 50 and over, there is no difference in sexual frequency between those who are divorced/widowed and those who have never married. You can see that Cramer's V is .021 and not significant. Thus, the relationship between marital status (excluding married individuals) and frequency of sex would appear to be spurious, produced by variations in age. Widowed and divorced persons do not have sex less frequently than do never-married people *in their age group!* The initial differences between the groups are the result of the fact that, as a group, people who are widowed or divorced are far more likely to be over 50 than are persons who have never married, most of whom are young adults who soon will marry. So our diagram is an accurate summary of the relationships.

These results are unusually clean. Typically, when control variables are used, there will be some "bounce," or random fluctuation, in the subtables because of the small numbers of cases in some columns. We can see more of the messiness of real analysis in the following example of an intervening variable.

Social scientists have noted that some individuals move because they are "pushed" by lack of opportunities in their current location, whereas others move because they are "pulled" by the attraction of opportunity elsewhere; this is called the "push/pull" hypothesis of geographic mobility. This hypothesis suggests that individuals who have moved since they were 16 would be likely to have higher incomes than those who didn't move. Let's check this. (Before continuing with the guide below, go to the variable selection screen and click on the [Clear All] button to clear all the previously used variables.)

| | | 63) MOVERS | |
		Stayer	Mover
76) INCOME	Under 15K	25.4%	18.8%
	15K-29,999	25.6%	21.7%
	30K & over	48.9%	59.5%
	TOTAL	100.0%	100.0%

Data File: **GSS96**
Task: **Cross-tabulation**
➤ *Row Variable:* **76) INCOME**
➤ *Column Variable:* **63) MOVERS**
➤ *View:* **Tables**
➤ *Display:* **Column %**

V=0.105**

Because this table has three rows, it is a little more difficult to read. However, we can make this table easier to read by collapsing the first two rows together—especially since the first two rows are very similar to one another. On the table itself, click on the row label *Under 15K* and it will be highlighted. Then click on the row label *15K–29,999* and it will be highlighted. Next click the [Collapse] button. Where it asks for a new category label, type **Under 30K** and click [OK]. The new table that is shown has the first two rows *temporarily* collapsed together. (Don't worry, you haven't permanently changed the data file—the next time you use this variable it will have three categories again.) This new table is easier to read, as shown here.

| | | 63) MOVERS | |
		Stayer	Mover
76) INCOME	Under 30K	51.1%	40.5%
	30K & over	48.9%	59.5%
	TOTAL	100.0%	100.0%

V=0.103**

Looking at the top row, we can see that stayers are somewhat more likely than movers to have incomes under $30,000. Thus, movers generally have higher incomes than stayers. Cramer's V = .103 and Prob. = .000.

We might speculate that educational attainment is an intervening variable: Individuals move to get more education, and higher educational attainment leads to higher-paying jobs. We can diagram this argument:

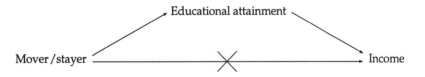

Educational attainment

Mover/stayer   Income

Notice that, when we are dealing with intervening variables, the variable comes between the other two variables in time. Again the arrows represent causal relationships. We have argued that educational attainment is an intervening variable between moving and income, so, in this situation, we expect the relationship between moving and income to disappear when we control for educational attainment. Let's check this idea using a version of the income variable that has been *permanently* collapsed into two categories, INCOME2. By using this second version of the variable, we can save ourselves the step of collapsing the first two rows manually at the end of our analysis.

Data File: **GSS96**	
Task: **Cross-tabulation**	
➤ *Row Variable:* **105) INCOME2**	
➤ *Column Variable:* **63) MOVERS**	
➤ *Control Variable:* **64) DEGREE**	
➤ *View:* **Tables (Not Hi Sch)**	
➤ *Display:* **Column %**	

Control: Not Hi Sch

	63) MOVERS	
105) INCOME2	Stayer	Mover
Under 30K	82.4%	72.5%
30K & over	17.6%	27.5%
TOTAL	100.0%	100.0%

V=0.115*

The table for those with less than a high school degree appears first. Among those who have not finished high school, movers are more likely than stayers to have higher incomes. The differences between stayers and movers (which is now 9.9 percent) is only slightly less than the difference between stayers and movers in the original table.

Data File: **GSS96**	
Task: **Cross-tabulation**	
Row Variable: **105) INCOME2**	
Column Variable: **63) MOVERS**	
Subset Variable: **64) DEGREE**	
➤ *View:* **Tables (High School)**	
➤ *Display:* **Column %**	

Control: High School

	63) MOVERS	
105) INCOME2	Stayer	Mover
Under 30K	51.9%	44.3%
30K & over	48.1%	55.7%
TOTAL	100.0%	100.0%

V=0.076**

If you are continuing from the previous analysis, simply click the appropriate button at the bottom of the task bar to look at the second (or "next") partial table for DEGREE (i.e., those with high school degrees).

Among high school graduates, the movers are likely to have higher incomes, but the difference (7.6 percent) is smaller than that in the original table. Now click the button at the bottom of the task bar to advance to the next table (those with some college).

Data File: **GSS96**

Task: **Cross-tabulation**

Row Variable: **105) INCOME2**

Column Variable: **63) MOVERS**

Subset Variable: **64) DEGREE**

➤ View: **Tables (Some College)**

➤ Display: **Column %**

Control: Some College

63) MOVERS		
	Stayer	Mover
105) Under 30K	24.9%	24.5%
INCOME2 30K & over	75.1%	75.5%
TOTAL	100.0%	100.0%

V=0.004

Among those with some college, there is no difference in income between the movers and stayers.

Since the original relationship has been weakened (or even disappeared) in these subtables, our hypothesis has been supported—educational attainment may be an intervening variable between moving and income. When we controlled for the intervening variable, the relationship between the other two variables disappeared or at least weakened. The diagram provides an accurate summary of the relationships.

Let's pause for a moment to summarize the difference between spurious and intervening relationships. To test for spuriousness, we control for a variable and examine the relationship between two variables. To test for an intervening variable, we do exactly the same thing. The only difference is the time order of the variables. With a spurious relationship, the control variable precedes the other two variables. With an intervening relationship, the control variable comes between the other two variables.

These concepts are slightly complex, so let's make sure they are understood before we go on. In the first example, there was a correlation between age and marital status, as well as a correlation between age and frequency of sex. But a third relationship between marital status and frequency of sex disappeared when you controlled for age. As the model showed, age preceded both marital status and the frequency of sex, and the apparent relationship between marital status and frequency of sex was spurious.

In the second example, there appeared to be a relationship between whether people moved from their home town since the age of 16 and their current level income. But one of the reasons that people move is to get an education—and people with more education typically have higher incomes. When we controlled for education, the original effect was greatly weakened (and in some cases disappeared). Thus, education was a link—an intervening variable—to the relationship between mover/stayer and one's current level of income.

Let's look at another example of a three-variable relationship. We might reason that individuals with low family incomes are more likely to live in neighborhoods with high crime rates and, consequently, might be more worried about walking alone at night. Let's check this.

Data File: **GSS96**

Task: **Cross-tabulation**

➤ Row Variable: **31) FEAR WALK**

➤ Column Variable: **76) INCOME**

➤ View: **Tables**

➤ Display: **Column %**

76) INCOME			
31) FEAR WALK	Under 15K	15K-29,999	30K & over
Yes	51.4%	43.3%	37.3%
No	48.6%	56.7%	62.7%
TOTAL	100.0%	100.0%	100.0%

V=0.113**

In fact, this appears to be the case. We can see that the higher the income category is, the less likely respondents are to say that they fear walking alone in their neighborhood at night.

But we may also think that crime is generally associated with cities. We might further think that people with low incomes are more likely to live in urban neighborhoods than are people with higher incomes. Hence, we might hypothesize that if we control for the urbanity of a neighborhood, the relationship between income and fear of walking alone in one's neighborhood at night will disappear.

In the following diagram, we have predicted that the relationship between income and fear of walking will disappear when we control for urbanity of neighborhood.

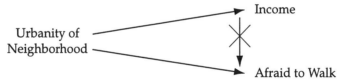

To test this idea, we would control for area of residence, 80) PLACE SIZE, and reexamine this relationship. So, repeat the previous analysis using PLACE SIZE as the control variable. You should be able to obtain each of the following tables (remember to use column percentaging for each control table).

Data File: **GSS96**

Task: **Cross-tabulation**

Row Variable: **31) FEAR WALK**

Column Variable: **76) INCOME**

➤ Control Variable: **80) PLACE SIZE**

➤ View: **Tables**

➤ Display: **Column %**

Control: City/subur

76) INCOME			
31) FEAR WALK	Under 15K	15K-29,999	30K & over
Yes	53.0%	47.6%	38.6%
No	47.0%	52.4%	61.4%
TOTAL	100.0%	100.0%	100.0%

V=0.121**

Control:Small city

| | | 76) INCOME | | |
		Under 15K	15K-29,999	30K & over
31) F E A R W A L K	Yes	56.3%	39.5%	41.7%
	No	43.8%	60.5%	58.3%
	TOTAL	100.0%	100.0%	100.0%

V=0.139

Control: Town/farm

| | | 76) INCOME | | |
		Under 15K	15K-29,999	30K & over
31) F E A R W A L K	Yes	41.1%	27.5%	28.8%
	No	58.9%	72.5%	71.2%
	TOTAL	100.0%	100.0%	100.0%

V=0.114

Study each of these tables. We can see that area of residence does have a powerful effect; in both the large and small city environments, over 50 percent of those in the lowest income group reported fear of walking at night, while only 41 percent of this group were afraid in town or farm areas in the country. However, the effect of income still persists, regardless of area of residence. Aside from some minor bounce, individuals with lower incomes reported more fear than individuals with higher incomes. So in this example, both variables have an effect on the dependent variable, and the diagram is *not* correct.

We could rediagram this result to reflect the observed relationships:

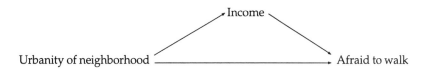

In this example, both independent variables have an effect on the dependent variable.

Your turn.

NAME: _____

COURSE: _____

DATE: _____

Workbook exercises and software are copyrighted. Copying is prohibited by law.

EXERCISE

5a

WORKSHEET

1. a. Let's use the 1993 GSS data to test some ideas. Before doing the cross-tabulation analysis in the following guide, look at the variable description for 51) READ PAPER and write the description below.

> ➤ *Data File:* **GSS93**
> ➤ *Task:* **Cross-tabulation**
> ➤ *Row Variable:* **23) VISIT ART**
> ➤ *Column Variable:* **51) READ PAPER**
> ➤ *View:* **Tables**
> ➤ *Display:* **Column %**

What is the description of variable 51) READ PAPER?

_____

_____

_____

b. Fill in the column percentages in the following table.

	Daily	Weekly	Seldom
Yes	_____%	_____%	_____%
No	_____%	_____%	_____%

V = _____

Prob. = _____

Is this significant? (Circle one.)   Yes   No

Describe the relationship shown in this table.

c. We might argue that this is a spurious relationship and that both of these variables are affected by educational attainment. Draw the diagram that would represent this argument.

d. Now conduct the same analysis using 48) DEGREE as the control variable. Fill in each of the tables using column percentaging, and write in Cramer's V and the significance level for each table.

> Data File   **GSS93**
> Task: **Cross-tabulation**
> Row Variable: **23) VISIT ART**
> Column Variable: **51) READ PAPER**
> ➤ Control Variable: **48) DEGREE**
> ➤ View: **Tables**
> ➤ Display: **Column %**

Control Category: NOT HI SCH

	Daily	Weekly	Seldom
Yes	_____%	_____%	_____%
No	_____%	_____%	_____%

V = _____

Prob. = _____

Is this significant? (Circle one.)   Yes   No

Control Category: HIGH SCH

	Daily	Weekly	Seldom
Yes	_____%	_____%	_____%
No	_____%	_____%	_____%

$$V = \underline{\hspace{2cm}}$$

$$Prob. = \underline{\hspace{2cm}}$$

Is this significant? (Circle one.)   Yes   No

Control Category: SOME COLL

	Daily	Weekly	Seldom
Yes	_____%	_____%	_____%
No	_____%	_____%	_____%

$$V = \underline{\hspace{2cm}}$$

$$Prob. = \underline{\hspace{2cm}}$$

Is this significant? (Circle one.)   Yes   No

e. For each of the following groups, indicate the percentage of respondents who said they visited an art museum in the last 12 months and who read newspapers daily.

Respondents without high school degrees   _____%

Respondents with high school degrees   _____%

Respondents with some college   _____%

f. Does the analysis support the diagram you drew in 1c?
(Circle one.)   Yes   No

g. In your own words, provide an interpretation of these results. Cite evidence from the tables to support your answer.

2. a. Fill in the column percentages in the table that follows. (Be sure to read the variable descriptions before conducting your analysis.)

Data File   **GSS93**
Task: **Cross-tabulation**
➤ Row Variable: **3) HUNT/FISH**
➤ Column Variable: **52) CH.ATTEND**
➤ View: **Tables**
➤ Display: **Column %**

	Not often	Often
Yes	_____%	_____%
No	_____%	_____%

V = _____

Prob. = _____

Is this significant? (Circle one.)   Yes   No

How would you interpret these results?

b. We might argue that this is a spurious relationship and that both of these variables are affected by gender: Women are more likely to go to church, and women are less likely to hunt. Draw the diagram that would represent this argument.

c. Conduct the same analysis using SEX as the control variable.

> Data File   **GSS93**
> Task: **Cross-tabulation**
> ➤ Row Variable: **23) VISIT ART**
> ➤ Column Variable: **52) CH.ATTEND**
> ➤ Control Variable: **39) SEX**
> ➤ View: **Tables**
> ➤ Display: **Column %**

Control Category: MALE

	Not often	Often
Yes	_____%	_____%
No	_____%	_____%

V = _____

Prob. = _____

Is this significant? (Circle one.)　　　　　　　　　　Yes　　No

Control Category: FEMALE

	Not often	Often
Yes	_____%	_____%
No	_____%	_____%

V = _____

Prob. = _____

Is this significant? (Circle one.)　　　　　　　　　　Yes　　No

d. Does the analysis support the diagram you drew in 2b?
(Circle one.)　　　　　　　　　　Yes　　No

e. In your own words, provide an interpretation of these results. Cite evidence from the tables to support your answer.

3. Let's test the following causal hypothesis: Veterans are less afraid to walk at night in their neighborhoods than are non-veterans.

Data File: **GSS93**
Task: **Cross-tabulation**
➤ Row Variable: **47) FEAR WALK**
➤ Column Variable: **35) VETERAN?**
➤ View: **Tables**
➤ Display: **Column %**

a. What is the description of variable 35) VETERAN?

_____

_____

b. Use column percentages to fill in the following table.

	Veteran	Non-veteran
Yes	_____%	_____%
No	_____%	_____%

V = _____

Prob. = _____

Is this significant? (Circle one.)   Yes   No

Do the results support or reject the hypothesis?
(Circle one).   Support   Reject

c. Select a variable in the GSS93 data set that might be a source of spuriousness. What is the number and name of this variable? (Hint: Avoid variables that are based on "attitudinal" survey questions. Also avoid variables that have more than two category labels.)

d. Why did you select this variable? Why do you think it may be a source of spuriousness?

e. Print each control table showing the column percentaging and attach printouts to this page. (If you have been instructed to not print out results, copy down the tables in the space below.) Then write down the value of Cramer's V and the significance level next to each table.

f. Draw the diagram that best represents these results.

g. Summarize your findings in a brief paragraph.

4. a. When we used the GSS96 data set earlier in this workbook, we found that some respondents expressed some impatience and hostility during the interview. Let's return to the GSS96 data file to see if we can gain some understanding of what other factors might affect this variable.

> *Data File:* **GSS96**
> *Task:* **Cross-tabulation**
> *Row Variable:* **33) ATTITUDE?**
> *Column Variable:* **64) DEGREE**
> *View:* **Tables**
> *Display:* **Column %**

Fill in the column percentages in the following table, and write in the values of Cramer's V and the significance level.

	Not hi sch	High sch	Some coll.
Friendly	_____%	_____%	_____%
Cooperative	_____%	_____%	_____%
Impat/hostile	_____%	_____%	_____%

V = _____

Prob. = _____

b. Does education appear to have an effect on degree of cooperation? If so, describe the relationship.

c. Perhaps education affects cooperation because individuals with lower educational attainment are more likely to fail to understand the questions and become frustrated. Perhaps degree of comprehension (COMPRE-HEND) is an intervening variable between educational attainment and cooperation. Show the diagram that would represent this argument.

d. Test this argument using 32) COMPREHEND and the cross-tabulation task. Finish labeling the following tables (to indicate the categories of the control variable) and fill in the results.

Control Category: _____

	Not hi sch	High sch	Some coll.
Friendly	_____%	_____%	_____%
Cooperative	_____%	_____%	_____%
Impat/hostile	_____%	_____%	_____%

V = _____

Prob. = _____

Control Category: _____

	Not hi sch	High sch	Some coll.
Friendly	_____%	_____%	_____%
Cooperative	_____%	_____%	_____%
Impat/hostile	_____%	_____%	_____%

V = _____

Prob. = _____

Control Category: _____

	Not hi sch	High sch	Some coll.
Friendly	_____%	_____%	_____%
Cooperative	_____%	_____%	_____%
Impat/hostile	_____%	_____%	_____%

V = _____

Prob. = _____

e. What is your interpretation of the results?

f. What practical implications does this result have for developing survey questions that will encourage cooperation?

5. Now let's use the ANES96 data file to test a proposition. First, let's look at the relationship between attitude toward welfare spending and whether the respondent has any college education.

> *Data File* **ANES96**
> *Task:* **Cross-tabulation**
> *Row Variable:* **53) WELFARE**
> *Column Variable:* **63) COLLEGE?**
> *View:* **Tables**
> *Display:* **Column %**

a. Write in the column percentages and the statistics below.

	No college	Some coll +
Decreased	_____%	_____%
Stay same	_____%	_____%
Increased	_____%	_____%

V = _____

Prob. = _____

Is this significant? (Circle one.)   Yes   No

b. Does having some college education appear to have an effect on attitude toward welfare spending? If so, describe the relationship.

c. Why would education level have any impact on attitudes toward welfare spending? Perhaps it's because people with higher educations have higher incomes—and people with higher incomes (and who pay higher taxes) may be more likely to think welfare spending should be decreased. In this case, income level would be an intervening variable between college education and attitude toward federal spending. That is, education level affects the income level of people and this in turn affects attitudes toward welfare spending. Draw a diagram to represent this proposition.

d. Perform the following analysis and fill in the results.

> Data File   **ANES96**
> Task: **Cross-tabulation**
> Row Variable: **53) WELFARE**
> Column Variable: **63) COLLEGE?**
> ➤ Control Variable: **62) INCOME CAT**
> ➤ View: **Tables**
> ➤ Display: **Column %**

Control Category: LOWER 50%

	No college	Some coll +
Decreased	_____%	_____%
Stay same	_____%	_____%
Increased	_____%	_____%

V = _____

Prob. = _____

Is this significant? (Circle one.)   Yes   No

Control Category: UPPER 50%

	No college	Some coll +
Decreased	_____%	_____%
Stay same	_____%	_____%
Increased	_____%	_____%

V = _____

Prob. = _____

Is this significant? (Circle one.)   Yes   No

e. What is your interpretation of these results?

5b

Causal Models

In the previous exercise, we saw how quickly cases can evaporate in cross-tabu-
lar analysis, making interpretations extremely difficult. This happens even
when we work with collapsed variables. In this exercise, we'll look at another
technique, regression analysis, that can be used for the same type of analysis. This
technique uses statistical procedures to unravel the influence of different independent variables and does not have the problem of running out of cases.

As you select the variables in the following guide, look at the variable
descriptions. The exclamation points in the names indicate that these are the
uncollapsed form of the variables and therefore these variables preserve the maximum amount of variation. In regression analysis, the more variation the better.
Notice that for 104) PRAY! in the GSS data file there are six categories of prayer
frequency ranging from *Never* to *More than once daily*. However, this is an ordinal
variable rather than being interval or ratio. Ideally, we use interval or ratio variables with regression, but we'll assume that these variables are close enough to
interval variables to be acceptable.

➤ *Data File:* **GSS96**
➤ *Task:* **Regression**
➤ *Dependent Variable:* **104) PRAY!**
➤ *Independent Variable:* **102) HEALTH!**
➤ *View:* **Graph**

We can see that people's perceptions of their own health seem to have some
effect on how often they pray. The worse people perceive their health to be, the more
frequently they pray. Looking at the value of multiple R-squared (R^2) which is
shown at the top right corner of the screen, we can see that health perceptions
explain 1 percent of the variation in frequency of praying (what appears on the
screen is the proportion of variance explained; to convert it into a percentage, simply

move the decimal point two places to the right). The effect of health perceptions is significant at the .01 level as the beta value of –0.102 is followed by two asterisks.

It's possible that health perceptions do not actually have a direct effect on prayer frequency. Age might have an impact on both health perceptions (since older people are in general likely to have worse health than younger people) and prayer frequency (since older people generally tend to be more committed to religion than younger people). Thus, it might be that age is an antecedent variable affecting both health perceptions and prayer frequency. Let's add 91) AGE! as a second independent variable in the regression analysis.

Data File:	**GSS96**
Task:	**Regression**
Dependent Variable:	**104) PRAY!**
➤ *Independent Variables:*	**102) HEALTH!**
	91) AGE!
➤ *View:*	**Graph**

Multiple R-Squared = 0.066**

HEALTH! ———— BETA = -0.057
(r = -0.102)

PRAY!

AGE! ———— BETA = 0.240**
(r = 0.251)

The correlation coefficient of each independent variable with the dependent variable is shown below each line. We can see that age is correlated with prayer frequency—older people pray more often than younger people. The R^2 figure shows that health perceptions and age combined explain 7 percent of the variation in prayer frequency.

The beta values indicate the effect of one independent variable when the effects of the other independent variables have been controlled. If we look at the beta for the effect of health perceptions (–.057), we can see that its effect has disappeared, and the beta coefficient is no longer significant. The beta coefficient for age (.240) shows that age does affect prayer frequency even when we control for the effects of health perceptions.

This regression diagram differs from those in Exercise 5a. Regression focuses on variation in the dependent variable and ignores relationships among the independent variables. Consequently, neither the time order of the two independent variables nor the relationship between them is represented in the diagram. We could show these same results using the type of diagram used in Exercise 5a.

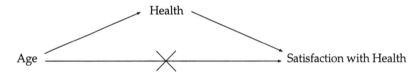

Let's look at the same information in a tabular form.

Data File: **GSS96**

Task: **Regression**

Dependent Variable: **104) PRAY!**

Independent Variables: **102) HEALTH!**

91) AGE!

➤ View: **ANOVA**

Analysis of Variance
Dependent Variable: PRAY!
N: 972   Missing: 1932
Multiple R-Square = 0.066   Y-Intercept = 3.686
Significance Levels: ** = .01, * = .05

Source	Sum of Squares	DF	Mean Square	F	Prob.
REGRESSION	146.457	2	73.229	34.161	0.000
RESIDUAL	2077.159	969	2.144		
TOTAL	2223.616	971			

	Unstand.b	Stand.Beta	Std.Err.b	t
HEALTH!	-0.109	-0.057	0.061	-1.789
AGE!	0.022	0.240	0.003	7.592 **

If you are continuing from the previous example, simply click the [ANOVA] option.

Only certain parts of this information are important for our purposes. Notice that the number of cases used in this analysis is only 972; all other cases had missing data on one or more of these variables. We can see that the effect of the two variables combined is statistically significant (Prob. = 0.000). The effect of 91) AGE! is significant at the .01 level; the *t* statistic is followed by two asterisks. The effect of 102) HEALTH! is no longer significant; the *t* statistic has no asterisks.

These results suggest that age is an antecedent variable affecting both health perceptions (and actual health) and prayer frequency. The effect of health perceptions on prayer frequency disappeared when we controlled for age.

In Exercise 5a, we looked at the effect of moving and education on income. Educational attainment appeared to be an intervening variable between moving and income. Let's look at this same example, using regression analysis. First, we should note that 63) MOVERS is not an ordinal variable; it is a nominal variable with two categories, 0 for stayers and 1 for movers. This is a *dummy* variable and can be used as an *independent* variable in regression analysis. Instead of using collapsed versions of the other variables as we did when we used tabular analysis, we'll use 95) INCOME! as our measure of income and 92) EDUCATION! as our measure of education.

Before you set up the regression analysis in the following guide, note that instead of individually deleting all the variables from the previous regression analysis, you can simply click the [Clear All] button.

Data File: **GSS96**

Task: **Regression**

➤ Dependent Variable: **95) INCOME!**

➤ Independent Variables: **63) MOVERS**

92) EDUCATION!

➤ View: **Graph**

Multiple R-Squared = 0.147**

MOVERS — BETA = 0.060** (r = 0.115)

INCOME!

EDUCATION! — BETA = 0.370** (r = 0.379)

The effect of moving on income has almost been completely removed by controlling for education. So again we can conclude that education is an intervening variable.

Now let's try this REGRESSION task with aggregate data. Use the USA data file. In Exercise 2a, we looked at the relationship between hunting and the murder rate and found a significant negative relationship. Murder is an interactive crime that almost always involves contact between the victim and the murderer. This would suggest that murder rates should increase when the amount of contact between individuals increases. For example, during summer months when the weather is pleasant, people spend more time outside and have more contact with one another. In fact, the murder rate is higher in summer than in winter. Areas also differ in their winter weather; in some states, winter is just as pleasant as summer. These states should have higher murder rates than states with extremely cold winters.

> *Data File:* **USA**
> *Task:* **Regression**
> *Dependent Variable:* **109) MURDER**
> *Independent Variables:* **72) HUNTING**
>   **37) WARM WINTR**
> *View:* **Graph**

Multiple R-Squared = 0.363**

HUNTING
BETA = -0.232
(r = -0.466)

WARM WINTR
BETA = 0.448**
(r = 0.569)

MURDER

The temperature in winter has a strong relationship with the murder rate: r = 0.57. Click on [ANOVA] to see the analysis of variance table. The effect of hunting is no longer significant. Controlling for weather caused a big drop in the effect of hunting on the murder rate. This suggests that the original negative relationship was spurious.

Your turn.

NAME: _____

COURSE: _____

DATE: _____

Workbook exercises and software are copyrighted. Copying is prohibited by law.

EXERCISE

5b

WORKSHEET

Note: In order to challenge you to think about independent and dependent variables, no Software Guides are included in the worksheets for Exercise 5b.

1. Using the GSS96 data file, test the following causal argument using regression analysis: "The higher their incomes, the more likely people are to support freedom of speech. However, this is really a spurious relationship, since both variables are the result of education—education increases both income and toleration of unpopular speech."

Diagram this argument using 92) EDUCATION!, 96) R.INCOME!, and 49) COMMUN SPK.

Use regression analysis to test this causal interpretation. Print the regression graphic. (Note: If your computer is not connected to a printer or if you have been instructed not to use the printer, just skip these printing instructions.)

a. What is the dependent variable?_____

b. List the independent variables, and provide the value of standard beta and the level of significance for each variable.

VARIABLE NAME	STANDARD BETA	SIGNIFICANCE (Circle one.)
_____	_____	Yes No
_____	_____	Yes No

c. What is the value of the multiple R^2? _____

d. What is the level of significance of R^2?   _____

e. Is the causal argument that education is the
   antecedent variable and the source of a spurious
   relationship supported? (Circle one.)   Yes   No

f. Cite the evidence which supports this conclusion.

2. Using the GSS96 data file, test the following causal argument using regression analysis: "Both father's and mother's education influence how much education their children get."

   Diagram this argument using 93) DAD EDUC!, 94) MOM EDUC!, and 92) EDUCATION!.

   Use regression analysis to test this causal interpretation. Print the regression graphic.

   a. What is the dependent variable?   _____

   b. List the independent variables, and provide the value of standard beta and the level of significance for each variable.

VARIABLE NAME	STANDARD BETA	SIGNIFICANCE (Circle one.)	
_____	_____	Yes	No
_____	_____	Yes	No

   c. What is the value of the multiple R^2?   _____

   d. What is the level of significance of R^2?   _____

e. Is the causal argument supported that the education
   of both parents causes the education of their offspring
   —that is, that each plays an independent causal role?
   (Circle one.)   Yes   No

f. Cite the evidence which supports this conclusion.

3. Now use 104) PRAY! as the dependent variable, 91) AGE! as the first inde-
    pendent variable, and 1) SEX as the second independent variable. Diagram
    the causal form you think these variables would fit. (Note: Sex is being
    treated as a dummy variable in this analysis, in which females are coded
    with a higher value than males. Hence, if the variable SEX is positively cor-
    related with another variable, it means that females are positively related
    with that variable. But if the relationship is negative, it means that females
    are negatively associated with that variable.)

Use regression analysis to test this causal interpretation. Print the regression
graphic.

a. List the independent variables, and provide the value of standard beta
   and the level of significance for each variable.

VARIABLE NAME   STANDARD BETA   SIGNIFICANCE (Circle one.)

_____   _____   Yes   No

_____   _____   Yes   No

b. What is the value of the multiple R^2?   _____

c. What is the level of significance of R^2?   _____

d. Interpret these findings.

4. Test the following causal argument using regression analysis: The strength of religious belief affects the frequency of praying. However, the link between strength of religious belief and frequency of prayer is commitment to conventional religious activities such as attendance at religious services.

    Diagram this argument.

    Use regression analysis to test this causal interpretation. (Hints: Use the search feature to find the variables you need. In selecting your variables, use _uncollapsed_ variables—variables that have an ! on the end of their names.) Print the resulting graphic.

    a. What is the dependent variable?   _____

    b. List the independent variables, and provide the value of standard beta and the level of significance for each variable.

VARIABLE NAME	STANDARD BETA	SIGNIFICANCE (Circle one.)	
_____	_____	Yes	No
_____	_____	Yes	No

c. What is the value of the multiple R^2? _____

d. What is the significance level of R^2? _____

e. Is the causal argument that conventional religious commitment (as represented by attendance at religious services) is
the intervening variable between strength of religious preference and frequency of praying? (Circle one.)   Yes   No

f. Explain your answer, citing the most relevant results.

5. Using the USA data file, look at the effect of 81) $ PER CAP. and 90) % COL-
LEGE on 100) BOOK $. Diagram what you believe to be the causal connec-
tions among these variables.

a. Explain your rationale for the causal relationship indicated above.

b. What is the dependent variable?   _____

c. List the independent variables, and provide the value of standard beta and the level of significance for each variable.

VARIABLE NAME   STANDARD BETA   SIGNIFICANCE (Circle one.)

_____   _____   Yes   No

_____   _____   Yes   No

d. What is the value of the multiple R^2?   _____

e. What is the level of significance of R^2?   _____

f. How would you interpret these results?

g. Was the causal relationship you predicted supported?
   (Circle one.)   Yes   No

6. Using the USA data file, look at the effect of 58) PLAYBOY and 27) MALE HOMES on 46) ALCOHOL.

a. What is the dependent variable?   _____

b. What is the value of r (shown below the line in the
   graphic) between 58) PLAYBOY and 46) ALCOHOL?   _____

c. What is the value of r between 27) MALE HOMES
   and 46) ALCOHOL?   _____

d. List the independent variables, and provide the value of standard beta and the level of significance for each variable.

VARIABLE NAME   STANDARD BETA   SIGNIFICANCE (Circle one.)

_____   _____   Yes   No

_____   _____   Yes   No

e. What is the value of the multiple R^2? _____

f. What is its level of significance? _____

g. How would you interpret these results?

Selecting a Study Design

OVERVIEW

Different kinds of research questions require different types of research designs. Each type of research (experiments, field studies, surveys, etc.) has advantages and disadvantages. In this exercise, you will gain experience in selecting the appropriate type of research in order to answer a particular type of research question.

BEFORE YOU BEGIN

Please make sure you have read Chapter 6 in the textbook and can answer the following review questions (you need not write any answers):

1. Describe the two fundamental features of experiments.

2. Explain why the experiment is the most powerful research design.

3. Give two reasons social scientists cannot always use experiments.

4. What is the big advantage of survey research, and what is its chief disadvantage?

5. What is field research, and in what situations is it most useful?

6. What is aggregate or comparative research, and how are data collected for such research?

7. Describe the general purpose of content analysis.

The text identifies the following types of research: survey research, comparative research, field research, experimental research, and content analysis. These categories reflect the way in which studies are generally described by social scientists: "Her experiment showed . . . ," "He used content analysis to test the idea that . . . ," "Our survey results suggest that . . . ," and so on. Classifying studies in this way identifies the type of research by its most salient element. Survey research, for example, emphasizes a data collection technique, while content analysis is basically a measurement technique, developing measures by coding content. This is a loose categorization and allows us to examine the elements that are usually combined in each design. In this exercise, we'll look at the types of research questions that are best suited to each design.

These designs differ in the degree of structure required in the research question. Experimental research requires a highly structured question, since the manipulation of the independent variable is built into the actual design of the study. Field research requires the least structured question since the researcher is able to observe anything of interest. The other approaches tend to fall between these two extremes. The formulation of the research question, then, is one of the first considerations in selecting a design.

In *exploratory* research, the research question is usually unstructured. Researchers are typically interested in a particular phenomenon and are still searching for connections with other phenomena. *Field* research is frequently used in this situation. For example, a researcher interested in how a new member is socialized into a deviant group might well start by observing how members of such a group interact with a new member or a potential member. When a research question is phrased using "how," field research is probably the best approach. Based on these observations, the researcher might develop some hypotheses on "why" certain approaches are more successful than others or "why" some members are more easily socialized than others.

When a researcher is interested in causes, or why certain phenomena happen, other designs are more appropriate. An *experiment* is the most powerful method for testing a causal hypothesis, since the problem of spuriousness is eliminated in the design of the study. So, when testing causal hypotheses, the researcher should first evaluate the possibility of conducting an experiment. Is the independent concept subject to manipulation? Can subjects be assigned randomly to levels of the independent variable? If the independent concept *can* be manipulated and subjects *can* be randomly assigned to levels of the independent variable, then the use of an experiment should be considered. In evaluating social programs, for example, subjects frequently can be assigned to different "treatments," and experimental research is a viable option. Incidentally, while most experiments use individuals as the unit of analysis, other units can be used. For example, a company selling to small businesses might use an experiment to compare the effectiveness of two different marketing approaches. In this study, small businesses would be the unit of analysis and each business would be assigned randomly to one of the two marketing approaches.

Two additional concerns should be addressed before deciding to use an experimental design: ethical questions and feasibility questions. Even if it is possible to manipulate the independent concept, it may not be ethical to do so. For example, one could use an experiment to assess the relative effectiveness of spanking in eliminating undesirable behaviors in children, but randomly assigning children to experimental treatments, one of which involved spanking, raises serious ethical questions. Even when the manipulation of the independent variable raises no ethical concerns, it may not be feasible to conduct an experiment. For example, one could test the effectiveness of a suicide-prevention education program by using an experiment: High school students could be randomly assigned to participate in the program or not. However, this study would not be feasible. Because suicide is a rare phenomenon and very few subjects in either treatment condition would ever commit suicide, an enormous number of subjects would be required to reliably compare the suicide rate across experimental conditions.

If an experiment cannot be used, naturally occurring variation of the independent concept must be measured. With the remaining three designs (survey research, comparative research, and content analysis), the logic of testing causal hypotheses is the same. With each, one must not only examine the relationship between the independent and dependent concepts, but also consider potential sources of spuriousness. So the choice of design hinges on the unit of analysis used in the research question and the nature of the concepts.

If the unit used in the research question is an aggregation, then the best test of the hypothesis would be with data on the relevant unit. With most aggregate units, the researcher can use only data that already has been collected—it wouldn't be feasible to collect data to determine the gross domestic product of each of a set of nations. Lack of appropriate data is frequently a problem in comparative research. With some aggregate units, surveys of "key" persons can be used to supplement existing data. For example, research questions about cities might be addressed by surveying mayors of an appropriate sample of cities. In some cases, it might be possible to use content analysis to create appropriate measures. For example, if you were interested in the effect of legal codes on various types of crime, you might be able to use content analysis of written legal codes to create the independent variable. Or with the Human Area Relations Files, you could use content analysis to code characteristics of societies from field notes recorded by anthropologists.

If the unit of analysis is an individual, then survey research is probably the most appropriate design. Except for ethical issues, the only limits on survey research are that you must be able to identify the appropriate population, to select a sample from this population, if necessary, and to collect information on the relevant concepts by asking questions.

Study designs are not mutually exclusive—elements of two or more may be incorporated into the same study. For example, in the experiment described in the text, subjects are shown information about a political candidate in which gender is the independent variable. Information on the dependent variable, candidate preference, could be collected in a variety of ways. The subjects might be asked to cast a single vote. Or they could be asked to fill out a questionnaire. Or they might be asked to write an essay on the relative merits of the candidates. Content analysis of the essay might be used to develop a measure of the dependent variable.

There are many other ways in which elements of more than one technique might be combined in a single study. A field researcher might survey the members of a group being observed. Or a survey study might have interviewers code information about an individual's living room—a bit of field research. Studies evaluating social programs frequently employ "field experiments"—randomly assigning subjects to one of several experimental treatments conducted in natural settings.

In designing research, the most important task is to examine the research question and to collect data in the manner most appropriate for answering this question.

Workbook exercises and software are copyrighted. Copying is prohibited by law.

1. For some research designs, it is very important to consider change over time. For each of the following research questions, circle **Yes** or **No** to indicate whether the research would need to use data from at least two different time periods.

   a. Is the "gender gap" in the political attitudes of men and women getting bigger or smaller? (Circle one.)   Yes   No

   b. Is the average family size smaller in more urbanized nations than it is in less urbanized nations? (Circle one.)   Yes   No

   c. As nations become more urbanized, does the average family size decrease? (Circle one.)   Yes   No

   d. Do members of the American Communist Party hold different views now than they did before the breakup of the Soviet Union? (Circle one.)   Yes   No

   e. Are Protestants and Catholics different from one another in their attitudes toward abortion? (Circle one.)   Yes   No

   f. Can a government program reduce drug abuse among young people? (Circle one.)   Yes   No

   g. Are regional differences in racial attitudes becoming smaller or larger? (Circle one.)   Yes   No

   h. Do younger people have different views toward material values than older people do? (Circle one.)   Yes   No

   i. Do young people today have different views on social issues than young people had twenty years ago? (Circle one.)   Yes   No

2. For each of the following research questions, circle **Yes** or **No** to indicate whether it is possible to use an experiment to answer the research question. If it is not possible to use an experiment, explain why. If an experiment is possible, describe any ethical issue or feasibility problem that might prohibit experimentation.

a. Do African Americans and white Americans hold different views on the issue of capital punishment?

NO (Explain below)

YES (Describe below any ethical issue or feasibility problem)

b. In the U.S. Senate, are Democratic senators more pro-labor than Republican senators are?

NO (Explain below)

YES (Describe below any ethical issue or feasibility problem)

c. Are juries more likely to convict an African-American person than a white person?

NO (Explain below)

YES (Describe below any ethical issue or feasibility problem)

d. Are people more likely to be persuaded by a printed message than a spoken message?

NO (Explain below)

YES (Describe below any ethical issue or feasibility problem)

e. Are sports fans more aggressive than people who are not sports fans?

NO (Explain below)

YES (Describe below any ethical issue or feasibility problem)

f. When election ballots list candidates alphabetically, does this give an advantage to people whose names start with letters that come early in the alphabet?

NO (Explain below)

YES (Describe below any ethical issue or feasibility problem)

3. For each of the following research questions, circle the design (experimental, survey, comparative, field, or content analysis) that you think is best and explain why.

a. Is *Newsweek* magazine more favorable to Democrats than to Republicans? (Circle one.)

   Experimental   Survey   Comparative   Field   Content Analysis

   Explain:

b. Are Jews more politically tolerant than Protestants and Catholics? (Circle one.)

   Experimental   Survey   Comparative   Field   Content Analysis

   Explain:

c. Do employers discriminate against short male applicants, preferring to hire tall men? (Circle one.)

Experimental   Survey   Comparative   Field   Content Analysis

Explain:

d. Do students get better grades the closer they sit to the front of the classroom? (Circle one.)

Experimental   Survey   Comparative   Field   Content Analysis

Explain:

e. Are people who listen to radio talk shows more conservative than those who don't? (Circle one.)

Experimental   Survey   Comparative   Field   Content Analysis

Explain:

f. What kinds of people stop and give money to people who stand at the curb with a sign saying they are homeless? (Circle one.)

Experimental   Survey   Comparative   Field   Content Analysis

Explain:

g. If an instructor writes encouraging notes to students on their first exams, does this increase student performance on the next exam? (Circle one).

    Experimental   Survey   Comparative   Field   Content Analysis

Explain:

h. What are similarities and differences between bartenders and barbers in terms of how they handle conversations with customers? (Circle one.)

    Experimental   Survey   Comparative   Field   Content Analysis

Explain:

4. Different kinds of research questions require different kinds of units of analysis (cases). For example, if you want to explain why some people are more supportive of democratic principles than others are, then you would use individuals as the units of analysis. For each of the following research questions, write in the blank space the unit of analysis for which you would need to collect data.

a. How is the degree of urbanism in states related to the crime rates in states?

_____

b. Are the political and social views of people related to the kinds of music they prefer?

_____

c. Do increases in economic development within nations lead to smaller family sizes?

_____

d. What kinds of newspapers are most likely to endorse Democratic candidates for public office?

_____

e. What kinds of social, economic, and political values are reflected in prime-time dramas on television?

_____

f. Does the size of a police department have an effect on how responsive it is to citizen complaints?

_____

g. Do bars tend to specialize? That is, do different bars attract different kinds of regular customers?

_____

h. Why are some families more likely to do things together than others are?

_____

5. Sometimes, we need to combine more than one method in order to investigate research questions. For each of the following situations, circle the *two* methods that we would probably need to use and provide a brief description of how they would be combined.

a. What kinds of people (in terms of personality, social attitudes, political views, and background characteristics) are most likely to be persuaded by an antismoking documentary? (Circle two.)

Experimental   Survey   Comparative   Field   Content Analysis

Briefly describe how you would combine these methods.

b. Are there liberal-conservative differences among news magazines (*Time*, *Newsweek*, etc.), and are these differences reflected in the views of their readers? (Circle two.)

Experimental   Survey   Comparative   Field   Content Analysis

Briefly describe how you would combine these methods.

c. Do the "messages" of restroom-wall graffiti vary according to the type of social environment in which the restroom exists? (Circle two.)

Experimental   Survey   Comparative   Field   Content Analysis

Briefly describe how you would combine these methods.

d. Are people who display flags outside their houses on the Fourth of July more supportive of basic democratic principles than are those who do not display flags? (Circle two.)

Experimental   Survey   Comparative   Field   Content Analysis

Briefly describe how you would combine these methods.

Survey Analysis

OVERVIEW

In this exercise, you'll learn about entering survey data into a computer file, checking for certain types of errors, deciding which responses (e.g., "don't know") to treat as missing, collapsing variables into a smaller set of categories to make them more convenient for cross-tabulation analysis, and recoding variables to create new variables. You will also gain experience in identifying problems in survey research.

BEFORE YOU BEGIN

Please make sure you have read Chapter 7 in the textbook and can answer the following review questions (you need not write any answers):

1. What three main factors cause reliability problems in survey research?

2. Compare the advantages/disadvantages of interviews and questionnaires.

3. Compare the advantages/disadvantages of telephone and face-to-face interviews.

4. What are the benefits of using standard questions in surveys?

5. Compare the advantages/disadvantages of closed and open-ended questions.

6. What problems (e.g., bias) should be considered in writing survey questions?

7. Why are forced option questions better than cafeteria questions?

8. What is a contingency question?

9. In terms of response bias among survey respondents, what are the problems of conformity and response set?

10. How is a trend study different from a panel or longitudinal study?

11. Describe the differences among age, cohort, and period effects.

12. What ethical concerns are particularly relevant to survey research?

For some of the following topics, the instructions will vary depending on whether you are using the DOS version or the Windows 95 version of Student MicroCase. Further, there are some features that are included only in the Windows 95 version. For these reasons, we will spend more time explaining the general concepts so that you will understand the basic ideas regardless of which version of the software you are using. Also, in some situations, we will refer you to "Creating a MicroCase File" at the beginning of the "Projects" section of this book. "Creating a MicroCase File" provides separate instructions for the DOS version and the Windows 95 version of Student MicroCase.

The primary challenge in conducting survey research is to ask questions that respondents can and will answer. The text provides considerable guidance in avoiding various problems in question construction, and you will get a chance to evaluate the quality of various questions in the written exercises. However, even after the data have been collected, there is still a great deal of work to do before you can analyze the data.

First, with the exception of some situations in which data are entered as the survey is being done (e.g., in computer assisted telephone interviews, or CATI), we need to enter the data from the survey into a computer file. Second, we need to prepare the data for analysis. This involves checking the accuracy of the data, identifying inappropriate responses, and eliminating unwanted categories. We might also need to *collapse* variables (e.g., collapsing education in actual years to a set of categories such as *under 8 years, 8–11 years*, etc.), and we might also need to *recode* variables to make them more useful for analysis (e.g., combining responses from several questions concerning participation in social activities to create a social participation index).

DATA ENTRY

The following illustrations show three questions from two hypothetical questionnaires filled out by two respondents. For each question, respondents were instructed to circle the number of the alternative that most reflected their answers. Thus, this questionnaire is *pre-coded*. That is, numbers—codes—have already been assigned to the question alternatives. If numbers had not already been assigned to the possible responses, then we would need to go through the questionnaires and code the responses.

Respondent 1

Sex: (1.) Male   2. Female

During the last twelve months, have you written a letter to any public official to express your views on any matter?

(1.)Yes   2. No   3. Don't Know

Please indicate your level of agreement or disagreement with the following statement: Judges should deal more harshly with criminals convicted of violent crimes.

1. Strongly Agree (2.)Agree 3. Disagree   4. Strongly Agree   5. Don't Know

---

Respondent 2

Sex: 1. Male   (2.)Female

During the last twelve months, have you written a letter to any public official to express your views on any matter?

1. Yes   2. No   (3.)Don't Know

Please indicate your level of agreement or disagreement with the following statement: Judges should deal more harshly with criminals convicted of violent crimes.

1. Strongly Agree   2. Agree   3. Disagree (4.)Strongly Agree   5. Don't Know

---

Next we need to organize the data prior to entering it into a computer file. The data are usually organized in a spreadsheet fashion with cases (respondents) along the side and variables across the top. For example, if we take the two hypothetical respondents and three hypothetical variables in the preceding illustrations, we would organize these data as follows:

	Variable 1 SEX	Variable 2 WRITE LET	Variable 3 HARSHER?
Respondent 1	1	1	2
Respondent 2	2	3	4

You have seen in the MicroCase data files that each variable has a number and a name. Here we named the first variable SEX, the second variable is WRITE LET, and the third variable is HARSHER?. Note that for SEX, the first respondent is coded 1 (for male). For WRITE LET, respondent 1 is coded 1 (for Yes). For HARSHER?, respondent 1 is coded 2 (for Agree). Respondent 2 is coded 2

(Female) for SEX, 3 (Don't Know) for WRITE LET, and 4 (Strongly Disagree) for HARSHER?.

Having organized the data, you would now be ready to create a data file and enter the data. As indicated before, the instructions for entering data are different for the two versions of Student MicroCase. Thus, at this point you might want to go to the appropriate section of "Creating a MicroCase File" at the beginning of the "Projects" section of this book and go through the brief tutorial on creating a data file. Alternately, your instructor might provide a separate data entry exercise for you to do.

Unfortunately, errors can occur in the process of gathering and entering data, and we will discuss ways of identifying certain types of errors in the next section. However, at the data entry stage, researchers frequently use *rekey verification* in order to minimize or eliminate the errors that can occur at this stage. Rekey verification means that the data are entered a second time and checked against the original data entries. If the second entry does not match the first entry, then the program notifies you so that you can see which entry is correct. This checking technique is based on the assumption that a data entry error is unlikely to occur a second time on the same response. You may remember that this is the same basic idea used to construct measures of reliability. By rekeying the data, we increase the reliability of the data.

To take a hypothetical example of rekey verification, when someone entered data for respondent number 800 for a religious preference variable, a 3 was entered. However, in doing rekey verification for this respondent for this variable, a 2 was entered. The rekey verification program will immediately respond that there is a discrepancy and ask which of the two values (the 3 or the 2) is correct. At this point, the person entering the data would determine which number is correct and select that number.

The Windows 95 version of Student MicroCase includes rekey verification, but the DOS version does not. For more information on rekey verification in the Windows 95 version, see the instructions in "Creating a MicroCase File" in the "Projects" section.

PREPARING DATA FOR ANALYSIS

Even after the survey data have been entered into a data file, there is still work to be done before we are ready to analyze the data.

CHECK DATA FOR ACCURACY AND INTEGRITY

After the data file has been created, we need to check the data entries to make sure that they are correct. Rekey verification (if it is available) can minimize errors at the data entry stage, but some data entry errors can still occur. Further, problems can occur prior to data entry. For example, a respondent's answer to a question might have been coded incorrectly. Another problem is that some respondents do not take the survey seriously and might give wrong answers as a joke or for other reasons. Such problems can occur regardless of whether the researcher creates the data file, obtains it from someone else, or imports the file from another program.

Beyond rekey verification, we can check data for accuracy by examining the data entries for certain kinds of problems. We begin this process by simply looking at the data to see whether everything looks right. Sometimes this can identify problems. In order to demonstrate this process, we have created a small fictitious TEST file. Let's look at the data in this file.

> ➤ *Data File:* **TEST**
> ➤ *Task:* **List Data**

Scroll across the screen so that you can view all the data. One problem that can occur is that one or more cases might not have the right values for variables entered in the proper columns. This can occur, for example, when the person entering the data gets off onto the wrong column—although MicroCase makes it very unlikely that errors of this sort will occur. Also, it is not unusual for something like this to happen when a researcher imports data from one program into another program. Thus, we first need to check to make sure that all the rows and columns end at the same point. If, for example, a row ended several columns before other rows did, then there is very likely a problem with the data for this case and it would need to be checked. As you can see, in the TEST file, all rows and columns end at the same point. Thus, the file looks good so far.

While all the rows and columns end at the appropriate points, this does not guarantee that the right data are in the right places. So, it would be a good idea to check some of the data for a couple of variables to make sure that the data look right. For example, we might check some of the cases for the last variable. If the last variable is correct, this indicates that the other variables are very likely lined up properly.

The next step is to examine the univariate statistics for each of the variables to determine whether anything looks like a mistake. Let's start with the first variable, SEX.

> *Data File:* **TEST**
> ➤ *Task:* **Univariate**
> ➤ *Primary Variable:* **1) SEX**
> ➤ *View:* **Bar - Freq.**

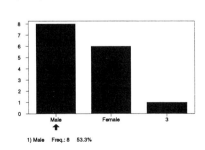

SEX -- Respondent's gender

1) Male   Freq.: 8   53.3%

Do you notice anything wrong here? Males are coded 1 and females are coded 2, but one case is coded 3. Thus, there is a data entry error, and we would need to check this case and correct the error. Since we have found one error for this case, it would be a good idea to check a few other variables for this case to make sure that this is just an isolated error rather than part of a pattern. Actually, this kind of data entry error would not occur in MicroCase because you can set the range for each categorical variable. Thus, the variable would have a low value of 1 and a high value of 2. If you tried to enter a 3 for this variable, the program would alert you that the value is outside the range of allowed values. Thus, using MicroCase, we had to "cheat" in order to even demonstrate this problem. However, not all statistical analysis systems have the capability to check the range.

We put in two more of these "errors" in this fictitious data file to demonstrate this kind of error. You might try finding them.

IDENTIFY INAPPROPRIATE RESPONSES

Sometimes, respondents give responses that are not accurate. For example, some survey respondents give joking responses that do not reflect their actual situations. We need to try to identify such responses, especially when they are extreme answers. There are several ways to check for such responses. When the values of a variable are not limited to particular categories, you should look at the univariate distribution to see if there are any extreme values. For example, in one year of the GSS, one individual claimed a lifetime total of 403 female sexual partners. This seemed highly unlikely when compared to the rest of the sample. But closer examination of this case revealed that the respondent was a 69-year-old man who reported having 5 to 10 sexual partners during the previous 12 months. He claimed casual pickups and paid sexual partners in addition to his wife. Assuming this man had been similarly active since the age of 20, his claim of 403 partners over his lifetime seems reasonable, even if the precision of the answer seems rather startling.

In the fictitious TEST file, let's look at the ATTEND variable.

Data File: **TEST**

Task: **Univariate**

➤ Primary Variable: **9) ATTEND**

➤ View: **Bar - Freq.**

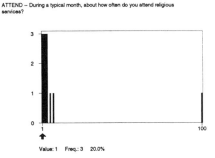

ATTEND -- During a typical month, about how often do you attend religious services?

Value: 1   Freq.: 3   20.0%

Note that for all but one of the respondents the answers given to this religious attendance question range from 1 to 8 times per month. The remaining respondent, however, indicated that he or she attended religious services 100 times per month. While this figure sounds very high and the roundness of the number makes it suspect, it is possible that it is actually correct. So, what do we do?

There are three options when we identify a suspicious value. First, we could change that value to missing data. Second, we could eliminate that case completely—if the respondent's answer to one question is a distortion, then the respondent's answers to other questions might also be unreliable. Third, we could create a "flag" variable. To use a "flag" variable, you first create a new variable with the values 0 and 1 (or actually any values that you wish). If a case appears to have inappropriate data, you set the value of this flag variable to 1; otherwise, the case is given a value of 0. When you analyze the data, you always select a subset of those cases that have a 0 on this flag variable. The advantage of using a flag variable rather than eliminating cases is that the data still exist for other analyses or other researchers.

Let's see how this flag variable option works. Note that the last variable in the TEST file is named FLAG VAR9—it is a flag variable for variable 9) ATTEND. For ATTEND, the only case that seemed unusual was the one that claimed to attend religious services 100 times a month. This case has been given a value of 1 for the flag variable and all other cases have a 0. So, we could use this flag variable to exclude any cases that are coded 1. Let's do this.

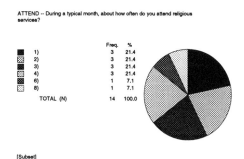

ATTEND -- During a typical month, about how often do you attend religious services?

		Freq.	%
■	1)	3	21.4
▨	2)	3	21.4
■	3)	3	21.4
▨	4)	3	21.4
■	6)	1	7.1
▨	8)	1	7.1
	TOTAL (N)	14	100.0

Data File: **TEST**
Task: **Univariate**
Primary Variable: **9) ATTEND**
➤ Subset Variable: **10) FLAG VAR9**
➤ Subset Categories: **Include: 0) Appropriat**
➤ View: **Pie**

[Subset]

Note that the exclusion of this case for this variable has quite an impact on the results. However, in a typical survey of 1,500 or so respondents, one highly unusual case such as this is not likely to have a great impact on the results.

Another way to check for inappropriate responses is to examine the associations between variables known to have certain relationships. For example, in a college survey, students may be asked their sex and their place of residence. We know that anyone living in a fraternity should be male. If we find one or more females claiming to live in a fraternity, we probably would want to check their responses to other questions. In a high school survey, one student claimed to be 7 feet tall and weigh 100 pounds. This is highly unlikely. Let's see how such checks might be done using the TEST file.

				7) REGISTERED		

Data File: **TEST**
➤ *Task:* **Cross-tabulation**
➤ *Row Variable:* **8) VOTE96**
➤ *Column Variable:* **7) REGISTERED**
➤ *View:* **Table**
➤ *Display:* **Column %**

7) REGISTERED			
8) VOTE96	Yes	No	Don't know
Clinton	41.7%	50.0%	0.0%
Dole	50.0%	0.0%	0.0%
Perot	8.3%	0.0%	0.0%
Other	0.0%	50.0%	0.0%
No Answer	0.0%	0.0%	100.0%
TOTAL	100.0%	100.0%	100.0%

V=0.870**

Note that two respondents who said they were not registered to vote in the 1996 elections also claimed to have voted—one for Clinton and one for Other. Thus, we know that for these two respondents, there are inappropriate responses somewhere. Either they were actually registered or they did not actually vote.

Eliminate Unwanted Categories

Note that the preceding cross-tabulation table (VOTE96 by REGISTERED) contained several categories that are not very useful for analysis in this situation. Another aspect of cleaning survey data for analysis involves answers such as "don't know," "other," and "no answer." Generally, we want these answers to be treated as missing data during analysis. In MicroCase, the easiest method of doing this is to list these as additional categories of missing data. Let's see how this works.

➤ *Data File:* **GSS96**
➤ *Task:* **Cross-tabulation**
➤ *Row Variable:* **56) EVER STRAY**
➤ *Column Variable:* **1) SEX**
➤ *View:* **Table**
➤ *Display:* **Column %**

1) SEX		
56) EVER STRAY	Male	Female
Yes	17.5%	11.1%
No	54.8%	65.7%
Never wed	25.7%	21.1%
No answer	2.0%	2.1%
TOTAL	100.0%	100.0%

V=0.120**

If you read across the row labeled "No answer," you will see that there are a total of 54 respondents who did not answer this question. In Exercise 5a of this workbook, you used a method to *temporarily combine* categories in a cross-tabulation table. This same technique can be used to *temporarily drop* a category. To do this, click on the category label "No answer" which causes the entire row to become highlighted. Then click on the [Collapse] button and choose the option to drop the category and turn the values to missing data. Click [OK] and the table returns to the screen. Notice that the "No answer" category has been temporarily removed. (If you were to repeat the cross-tabulation of these same two variables, the "no answer" category would reappear again.)

Student MicroCase has a second way to temporarily turn categories to missing data. Let's say you have a data set that has a large number of variables that include a "No answer" category—or some other unwanted category, such as "Don't know" or "No opinion." Student MicroCase allows you to designate a set of such categories so that whenever they are encountered during analysis they will automatically be turned to missing data. Here's how it works. Return to the main menu. If you are using the Windows 95 version of MicroCase, go to the FILE & DATA MENU and select the FILE SETTINGS option. (If you are using the DOS version of the software, go to the main menu and click the button "Set Categories to Missing Data.") You can enter up to six missing data categories that will be excluded from all subsequent analyses while using this data file. Go ahead and type **NO ANSWER** in the first blank. This category will now be excluded not only from EVER STRAY, but from all variables in all analyses that have this category in the GSS96 file. You could also enter DON'T KNOW or NO OPINION and these categories would be excluded from our analyses.

If we later change our minds and want to include any of these categories again, we can go back to this option and delete the category from the list. In your student version of the software, all categories you enter here will be erased when you open a different data file or exit the program.

Now that you have "No answer" assigned as a missing data category, return to the CROSS-TABULATION task and repeat your analysis of EVER STRAY and SEX.

Data File:	**GSS96**
➤ Task:	**Cross-tabulation**
➤ Row Variable:	**56) EVER STRAY**
➤ Column Variable:	**1) SEX**
➤ View:	**Table**
➤ Display:	**Column %**

1) SEX

56) EVER STRAY	Male	Female
Yes	17.8%	11.3%
No	55.9%	67.1%
Never wed	26.3%	21.6%
TOTAL	100.0%	100.0%

V=0.121**

Sure enough, the "No answer" category was automatically turned to missing data.

COLLAPSING VARIABLES

Previously in this workbook, you saw the usefulness of using collapsed versions of a variable. You may recall, for example, that all the variables in the data set that end with a "!" are the original, uncollapsed version of that variable. Here is the univariate distribution of the original age variable in GSS96:

Data File: **GSS96**
➤ Task: **Univariate**
➤ Primary Variable: **91) AGE!**
➤ View: **Bar - Freq.**

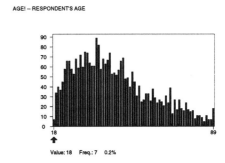

Notice that the original age categories range from 18 to 89. If we tried to use this variable in a cross-tabulation analysis, the resulting table would be nearly impossible to interpret. If you look at variable 66) AGE, you'll see that this variable has been collapsed into five categories.

Data File: **GSS96**
Task: **Univariate**
➤ Primary Variable: **66) AGE**
➤ View: **Bar - Freq.**

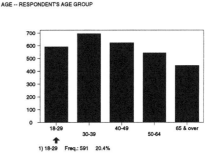

You may have wondered how the collapsed version of age was created. Did someone go through the data set and assign each case into one of these five categories and then key in the data for the new variable? In the old days that's how this was done. But today there are powerful programs like MicroCase that make this an easy procedure. Unfortunately, the task that is about to be described is available only in the Windows 95 version of Student MicroCase. So, if you are not using this version of the program, skip ahead to the next section, titled "Recoding Variables."

If you are still reading this paragraph, that means you are using the Windows 95 version of MicroCase. Go to the FILE & DATA MENU and select the COLLAPSE VARIABLE task. You are prompted to select a variable to collapse. Even though a collapsed version of AGE! already exists in this data set, let's create another one ourselves. So select **91) AGE!** as the variable to be collapsed, and click [OK]. This bar graph is similar to what we saw moments ago when we used the Univariate task to look at this variable. But in this task, we are able to combine specified values on the bar graph. In the lower, middle section of the screen is a list of buttons for categories of the new variable we are about to create. Click on Category 1 and type in **18–29**. Now position your mouse over the first category in the upper bar graph, which represents the respondents who are 18. Notice that there are 7 cases (or 0.2%) in this category (these values are shown below the bar graph). Click on this category to assign it to the new 18–29 category you created below. Click on the next 11 bars until all respondents coded in the 18–29 range have been assigned to this 18–29 category.

Now click on the Category 2 button in the list below and type **30–39**. Return to the upper bar graph and click on the 10 bars representing those respondents who are 30–39 so that they will be assigned to the 30–39 category you created. For Category 3 type in **40–49** and assign the appropriate categories from the graph above to this new category. You're probably getting the hang of this by now. Follow the same procedure for those who are 50–64 and for those who are 65 & OVER.

Recall that in the previous section we temporarily dropped categories from our analysis. However, if we wanted to permanently drop one of these categories, we could do so by simply not assigning it to one of the categories below. Any value in the original variable that is not assigned to a category on the new variable will become missing data and will be automatically excluded from analyses when the new variable is used. In this example, we wanted to assign all of the categories from AGE! to our new variable, so all the categories have been assigned.

As you probably noticed, as we collapsed variables, the horizontal bar graph in the lower right kept a running total of the size of each new category. This helps us balance the sizes of categories if that's an important consideration. Note that we could click the [Cumulative] button on the left if we wanted to work with a cumulative bar graph instead of a regular bar graph. A cumulative graph is often helpful with interval/ratio or ordinal variables in assigning given proportions of cases to new categories. For example, we might want to divide the age distribution into thirds.

We are nearly finished creating our new variable. Click the [Finish] button and a final window will appear on the screen. MicroCase's COLLAPSE VARIABLE option does not replace the original variable but, rather, creates a new variable and places it at the end of the data file. At this point we can enter a name for our new variable. MicroCase suggests just adding a 2 to the old variable name—AGE!2. Let's instead replace that suggestion with the name AGE CATEG, which will remind us that this is an age variable that has been collapsed into categories. If you want to modify your variable description, you can do that as well. Again, MicroCase has suggested some wording for the new variable description, but you may want to modify this to your liking. Click [OK] and you will be told that a new variable has been added to your GSS96 data file. If you click **[OK]** again, you will be returned to the main menu. The AGE CATEG variable will remain for you to use unless you replace it with another variable. Student MicroCase allows you to add 10 variables like this to your data file. If necessary, new collapsed variables can replace any of these 10 variables.

Let's take a look at the new variable we created.

<table>
<tr><td align="right">Data File:</td><td>GSS96</td></tr>
<tr><td align="right">Task:</td><td>Univariate</td></tr>
<tr><td align="right">➤ Primary Variable:</td><td>AGE CATEG</td></tr>
<tr><td align="right">➤ View:</td><td>Bar - Freq.</td></tr>
</table>

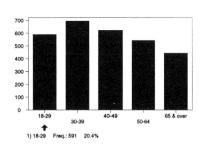

Sure enough. This looks identical to the 66) AGE variable that had already been collapsed in the GSS96 data file.

Recoding Variables

The collapse option discussed in the previous section is actually one of many methods for creating *recoded variables*. Recoded variables are transformations or computations made to the original set of variables in order to create entirely new variables. There are many different types of recodes used in survey research. One of the more common recodes is to create an *index* that combines responses from two or more questions.

<table>
<tr><td align="right">Data File:</td><td>GSS96</td></tr>
<tr><td align="right">Task:</td><td>Univariate</td></tr>
<tr><td align="right">➤ Primary Variable:</td><td>47) ATHEIST SP</td></tr>
<tr><td align="right">➤ View:</td><td>Pie</td></tr>
</table>

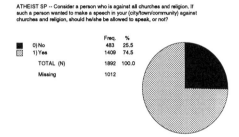

Notice that this variable is coded 0 for respondents who would not allow free speech in this situation and it is coded 1 for those who would allow free speech. If you examine the variable descriptions for the next four variables in the data set, 48–51, you will see that the same coding scheme (0=No; 1=Yes) is used. We could compute a new variable that is based on the summing of the number of "allow speech" responses that each respondent gave to the five freedom of speech questions. In fact, variable 52) FREE SPEAK has done just that.

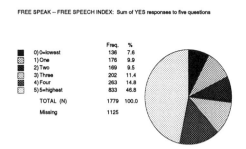

Data File: **GSS96**

Task: **Univariate**

➤ Primary Variable: **52) FREE SPEAK**

➤ View: **Pie**

FREE SPEAK -- FREE SPEECH INDEX: Sum of YES responses to five questions

		Freq.	%
■	0) 0=lowest	136	7.6
▨	1) One	176	9.9
■	2) Two	169	9.5
▨	3) Three	202	11.4
▨	4) Four	263	14.8
▨	5) 5=highest	833	46.8
	TOTAL (N)	1779	100.0
	Missing	1125	

As you can see, those respondents who answered Yes to each of the five free speech questions are coded as 5. Those respondents who answered Yes to all but one of the questions, are coded as 4, and so on. (If a respondent did not provide an answer for each of the five questions, they were assigned to missing data for this variable.)

Once again your curiosity as to how such an index is computed is probably overwhelming you. If you are using the Windows 95 version of Student MicroCase, your curiosity will be satisfied. If you are not using the Windows 95 version of Student MicroCase, skip ahead to the worksheet section.

With the Windows 95 version of Student MicroCase, go to the FILE & DATA MENU and select the RECODE VARIABLES task. Then select the [Sum/Index] option. The screen that appears gives you numerous choices for the creation of your recode. Although you are welcome to read the various options, do not change the original, default settings for this example. Click the [Next] button to continue. Now select variables **47–51** as the variables to be included in the sum recode. Click the [Next] button and you are presented with a screen that lets you evaluate the reliability of the index you are creating. At the lower part of the screen, you are shown the correlations between each of the variables you selected for the index. At the top of the screen is Cronbach's alpha. You will recall from Chapter 3 in the textbook that Cronbach's alpha is a popular measure of reliability that is based on the correlation—the internal consistency—of the variables that comprise an index. The rule of thumb is that if Cronbach's alpha is over .70, the measure is sufficiently reliable. Since the value we obtained is .82, we seem to be on safe ground.

Click the [Next] button and you are prompted for the variable name of the index you are creating. Let's call this variable **SPEAK INDX**. If you would like to modify the variable description that MicroCase has automatically generated, you can do that at this time too. Before you continue, take a quick look at the matrix provided at the bottom of this screen. The first column, shown in red, lists the individual values for the new variable you are creating. The next five columns show the coding for the original five variables selected for the index. Examine the first case. Notice that it is coded as 1 for the new variable. If you scroll across and look at the coding of this case on the other five variables, you will see, as expected, that this person answered Yes (coded as 1) to only one of the free speech questions. If you go back and examine the third case, you'll see that it is coded as 5 in our new variable and that this person answered Yes to each of the five free speech questions.

After you are done examining this screen, click [Next] to return to the final screen. Here you are simply reminded that your new variable, SPEAK INDX, will be added to the end of the data file. Click [Finish] and [OK] to return to the main menu.

Let's look at the final results of the variable we created.

Data File: **GSS96**

Task: **Univariate**

➤ *Primary Variable:* **SPEAK INDX**

➤ *View:* **Pie**

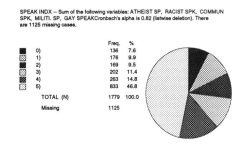

SPEAK INDX -- Sum of the following variables: ATHEIST SP, RACIST SPK, COMMUN SPK, MILITI. SP, GAY SPEAK Cronbach's alpha is 0.82 (listwise deletion). There are 1125 missing cases.

	Freq.	%
0)	136	7.6
1)	176	9.9
2)	169	9.5
3)	202	11.4
4)	263	14.8
5)	833	46.8
TOTAL (N)	1779	100.0
Missing	1125	

No surprises here. The results match the 52) FREE SPEAK variable we looked at earlier.

Workbook exercises and software are copyrighted. Copying is prohibited by law.

NAME:

COURSE:

DATE:

1. Let's begin by comparing two variables from the ANES96 file.

> Data File: **ANES96**
> Task: **Cross-tabulation**
> Row Variable: **35) WHO VOTE?**
> Column Variable: **25) WHO VOTE**
> View: **Tables**
> Display: **Column %**

The 1996 American National Election Studies (ANES) study surveyed people before the presidential election and then again after the election was over. Variable 25) WHO VOTE was part of the pre-election survey in September; it asked about the respondent's choice for president in the upcoming election. After the election, variable 35) WHO VOTE? asked which candidate the respondent had voted for in the 1996 presidential election. Given these two variables, we can compare respondents' pre-election preferences with their reported vote. Write in the column percentages in the table below.

	Clinton	Dole	Perot
Clinton	_____	_____	_____
Dole	_____	_____	_____
Perot	_____	_____	_____

a. Discuss how well we would be able to predict the final voting choice of the respondents on the basis of their pre-election preferences.

b. In November's presidential election, some people obviously did not vote in accord with the pre-election preferences they had in the September survey. There are at least two possible reasons for discrepancies in this preference between time 1 (pre-election) and time 2 (post-election). What two major reasons might account for these discrepancies?

2. Surveys are extremely useful in social research, but we need to interpret the results with some caution. Let's look at reported voting participation in the 1996 elections.

> Data File: **ANES96**
> ➤ Task: **Univariate**
> ➤ Primary Variable: **34) VOTED?**
> ➤ View: **Pie**

a. Write the variable description for this voting-participation question below.

_____

_____

_____

_____

_____

b. Notice the wording of this question. Why didn't the researchers simply ask: "Did you vote in the 1996 election?" What problem were the researchers trying to minimize by asking this voting-participation question the way they did?

c. According to the univariate statistics, what percentage of those who answered this question claimed that they voted in the 1996 elections?   _____%

d. We know from voting statistics that only about 50 percent of the electorate voted for president in the 1996 elections. What might account for the difference between this figure and the percentage who claimed they voted?

3. Let's look at 35) WHO VOTE? again.

>   Data File: **ANES96**
>   Task: **Univariate**
> ➤ Primary Variable: **35) WHO VOTE?**
>   ➤ View: **Pie**

a. What percentage of the respondents said that they voted for Clinton?   _____%

b. On the basis of voting statistics, we know that President Clinton received just under 50 percent of the votes in 1996. Thus, there is a discrepancy between the percentage who say they voted for him and the actual percentage. There are at least two basic explanations for this discrepancy. One is sampling error. What is another possible explanation?

4. Based on what you learned from the textbook, discuss the problems with each of the following questions.

a. *Question: Do you think that high schools should or should not be prohibited from banning the wearing of gang colors by students?* (Circle one.)   *Should   Should not*

What is the major problem with this question?

How could this question be rewritten to eliminate the problem?

b. *Question: Should judges be allowed to continue coddling criminals by giving some of them suspended sentences for the first offense?* (Circle one.)   *Yes   No*

What is the major problem with this question?

How could this question be rewritten to eliminate the problem?

c. *Question: What is your view concerning abortion?* (Circle one.)

   A. *It should not be allowed.*
   B. *It should be allowed.*

What is the major problem with this question?

How could this question be rewritten to eliminate the problem?

d. *Question: How many times have you changed your major since you began college?* (Circle one.)

0–2 times   3–5 times   More than 5 times

What is the major problem with this question?

How could this question be rewritten to eliminate the problem?

e. *Question: During the last twelve months, have you done any of the following?* (Circle all that apply.)

Visited an art museum?
Attended a professional sports event?
Gone to a play at a theater?
Gone to an auto race?
Gone to an amateur sports event?

What is the major problem with this question?

How could this question be rewritten to eliminate the problem?

5. Open the GSS96 data set and do the following analysis.

➤ *Data File:*   **GSS96**
➤ *Task:*   **Univariate**
➤ *Primary Variable:*   **56) EVER STRAY**
➤ *View:*   **Pie**

a. What is the variable description of 56) EVER STRAY?

_____

_____

_____

b. Fill in the category labels and percentages for this variable.

　　　　　CATEGORY LABEL　　　　　　　　　PERCENTAGE

1. _____　　　_____

2. _____　　　_____

3. _____　　　_____

5. _____　　　_____

c. Which category should be treated as missing data in any analysis?

_____

d. Why are those who have never married placed in a separate category?

e. In the next two examples, we want to exclude (i.e., turn to missing data) certain categories from the variables we select. But rather than use the collapse or recode techniques described in the preliminary part of this exercise, we are going to use the subset option to achieve the same results.

Let's look at the relationship between 56) EVER STRAY and age for all respondents who answered the question about straying. Use column percentages and fill in the following table:

> | | |
> |---:|:---|
> | Data File: | **GSS96** |
> | ➤ Task: | **Cross-tabulation** |
> | ➤ Row Variable: | **56) EVER STRAY** |
> | ➤ Column Variable: | **67) OVER 50** |
> | ➤ Subset Variable: | **56) EVER STRAY** |
> | ➤ Subset Categories: | **Exclude 5) No Answer** |
> | ➤ View: | **Tables** |
> | ➤ Display: | **Column %** |

Remember, subset variables are selected at the same time that other variables are selected. In this example, make sure you EXCLUDE the No Answer category.

	Under 50	50 & Over
Yes	_____	_____
No	_____	_____
Never Wed	_____	_____

V = _____

Prob. = _____

f. Now let's exclude both the No Answer responses and the Never Wed responses. Then use column percentages to fill in the following table:

Data File:	**GSS96**
Task:	**Cross-tabulation**
Row Variable:	**56) EVER STRAY**
Column Variable:	**67) OVER 50**
Subset Variable:	**56) EVER STRAY**
➤ *Subset Categories:*	**Exclude 5) No Answer**
	3) Never Wed
➤ *View:*	**Tables**
➤ *Display:*	**Column %**

Perhaps the easiest way to modify the category selection for a subset variable is to delete the subset variable entirely and then select it again. For this example, make sure you EXCLUDE the No Answer and Never Wed categories.

	Under 50	50 & Over
Yes	_____	_____
No	_____	_____

V = _____

Prob. = _____

g. Explain which table would be more appropriate for testing the hypothesis that individuals 50 and over are more likely to have strayed than are individuals under 50.

h. How was the strength of the relationship affected by which categories were treated as missing data?

6. a. Let's go to the GSS93 data file and look at 37) PUB.DECIDE. Fill in the distribution below.

> ➤ Data File: **GSS93**
> ➤ Task: **Univariate**
> ➤ Primary Variable: **37) PUB.DECIDE**
> ➤ View: **Pie**

	CATEGORY LABEL	PERCENTAGE
1.	_____	_____
2.	_____	_____
8.	_____	_____
9.	_____	_____

b. Which category should be treated as missing data in any analysis? _____

c. The category "Can't Choose" could be treated either as missing data or as a middle category between categories 1 and 2. Which option would you choose and why?

7. Sometimes the elimination of missing data can have a sizable impact on the results, either increasing the strength of the relationship or decreasing the strength.

a. Look at the relationship between 37) PUB.DECIDE and 38) BUS.DECIDE. Use column percentaging to fill in the following table. Then find Cramer's V and the significance level using the statistics summary option.

> *Data File:* **GSS93**
> ➤ *Task:* **Cross-tabulation**
> ➤ *Row Variable:* **37) PUB.DECIDE**
> ➤ *Column Variable:* **38) BUS.DECIDE**
> ➤ *View:* **Tables**
> ➤ *Display:* **Column %**

	Business D	Gov.Laws	Can't Chos	No Answer
People Dec	_____	_____	_____	_____
Gov. Laws	_____	_____	_____	_____
Can't Chos	_____	_____	_____	_____
No Answer	_____	_____	_____	_____

V = _____

Prob. = _____

b. Let's temporarily eliminate the *Can't Choose* and *No Answer* categories from both variables. First, click the *Can't Chos* label across the top of the table. Then click the *No Answer* label across the top of the table. These two columns are now highlighted. Next click the [Collapse] button and convert these categories to missing data (drop them). Then proceed to do the same thing with the *Can't Chos* and *No Answer* categories along the left side of the table. Now fill in the new column percentages in the following table and find Cramer's V and the significance level.

	Business D	Gov.Laws
People Dec	_____	_____
Gov. Laws	_____	_____

V = _____

Prob. = _____

c. How has the removal of these "missing data" categories affected the relationship?

8. The following paragraph describes a study conducted in 1935:

Long and Share Our Wealth stimulated the first scientific public opinion poll on a Presidential race. . . . Emil Hurja, the chief statistician and an executive director of the Democratic National Committee, mailed straw ballots and a cover letter on April 30, 1935 to about 150,000 people in the U.S. states. He made the poll appear as if it were being conducted by a magazine, the nonexistent National Inquirer. All the information was to be received from a three-by-five inch card—the two-cents postage was prepaid.[1]

---

[1] Edwin Amenta, Kathleen Dunleavy, and Mary Bernstein, "Huey Long's 'Share Our Wealth' and the Second New Deal," *American Sociological Review*, Oct. 1994.

What element of this study might raise ethical questions? Explain.

9. If you are using the Windows 95 version of Student MicroCase, let's compute an index. Note that variable 46) ABORT INDX in the GSS96 data file is an index based on the number of "allow abortion" responses given to six questions asking about abortion in various situations. Let's compute an index based on just three of these questions—those that concern emergency situations. The questions ask whether the respondent agrees that abortion is acceptable when the health of the mother in endangered, when there might be serious damage to the fetus, or when the pregnancy occurred because of rape. (If you have difficulty using this task, refer back to the last part of the preliminary section of this exercise.)

> *Data File:* **GSS96**
> *Task:* **Recode Variables (located on the FILE & DATA MENU)**
> *Recode Option:* **Sum/Index**
> *Select Variables:* **40) ABORT DEF**
> **42) ABORT HLTH**
> **44) ABORT RAPE**

After you have selected these variables for the index, click on [Next].

a. What is the value for Cronbach's alpha? _____

b. According to the textbook, is this value high enough to be reliable? (Circle one.)   Yes   No

Click [Next] again. The program presents the value for this new variable for each respondent. Note also that there are many missing values for this variable—because not all the respondents were asked this series of questions and some of those who were asked did not answer.

Scroll through the cases and write in the value of this new variable for the following cases. (Note: Write in *Missing* for a case if it has a blank.)

Case Number 1   _____

Case Number 28   _____

Case Number 31   _____

Case Number 42   _____

Now type a name for this new variable—let's call it **ABORT EMER**—and press <ENTER>. This index is complete.

Go to the Univariate task and examine the variable you created.

> Data File: **GSS96**
> ➤ Task: **Univariate**
> ➤ Primary Variable: **ABORT EMER**
> ➤ View: **Pie**

Print out this result and turn it in with your assignment.

Comparative Methods

OVERVIEW

In this exercise, you will first learn more about selecting the proper base for the calculation of rates (e.g., average yearly food stamp benefit per recipient) in comparative research. Because comparative research often uses a relatively small number of cases, each individual case might have a substantial effect on the results. Thus, you will see the effect that outliers can have on correlations and how this problem can be handled. Lastly, you will see that missing data can be a major problem in comparative research.

BEFORE YOU BEGIN

Please make sure you have read Chapter 8 in the textbook and can answer the following review questions (you need not write any answers):

1. In comparative research, describe what a rate is and give examples of rates that have different kinds of bases.

2. Compare the reliability of aggregate data with that of survey data, and explain why there is a difference.

3. What are the chief limitations of aggregate data based on official records?

4. If a rate such as a crime rate is not accurate, how can it still be useful for comparative research?

5. What is the difference between a case-oriented approach and variable-oriented research?

6. What are outliers, and why is it important to check for their existence in comparative research results?

7.   Why shouldn't you use 20 independent variables in a regression analysis to explain variation in welfare spending among the United States?

8.   What are the primary kinds of circumstances in which ethical issues need to be considered in comparative research?

In this exercise, we'll explore some of the special problems of comparative research. First, choosing an appropriate base (or denominator) for a rate can be difficult, and different bases may lead to quite different results. Second, there are usually a limited number of cases in the data file, so the impact of each case is considerably greater than in survey research. Finally, there may be a sizable amount of missing data.

In comparative research, almost all the variables are rates of one type or another. In selecting the base (or denominator), you always should consider the concept you are trying to measure. In some situations, changing the base helps improve the validity of the measure. For example, dividing the number of students by the *school-age* population would be a better measure of current participation in the educational system than would dividing the number of students by the *total* population. If we used the total population to create this rate, then a country with a small number of school-age children would have a low rate, even if every child was enrolled in school. The difference between the crude birth rate and the fertility rate described in the textbook is another example of refining a rate by changing the base.

Sometimes, however, changing the base may completely change the concept being measured. For example, if you divide the number of female students by the number of females in the school-age population, you will obtain the proportion of *school-age females* who are enrolled in school. On the other hand, if you divide the number of female students by the total number of students, you will get the proportion of *students* who are female. Clearly, these are not the same, and countries would not necessarily rank in the same order on the two variables. Countries in which only the elite enrolled in school but that had no sex discrimination would have a relatively low proportion of school-age females enrolled in school, but a relatively high proportion of students would be female.

Rates may change (or differ) because the values of either the denominator or the numerator or both change. Consider, for example, the percentage of deaths between ages 15 and 25 attributable to suicide. The numerator is the number of deaths by suicide in this age group, and the denominator is the total number of deaths in this age group. The rate can increase because the number of suicides increases, or because the number of deaths from other causes decreases, or both. If we are interested in the propensity of this age group to commit suicide, a better base would be the number of persons between ages 15 and 25.

Keep in mind that very small numbers and very large numbers are difficult to read and interpret. Consequently, most rates are adjusted so that the values fall in the range from 0 to 100 or multiples of 100. For example, rather than calculate the proportion of the population that was murdered (which would give extremely small numbers), we use the number of murders per 100,000 population. This is just the proportion murdered multiplied by 100,000. On the other hand, reporting the number of males per 100,000 population would create undesirably large numbers, so the percentage male—the proportion multiplied by 100—is usually used.

Remember that multiplying (or dividing) a rate by a constant value has no effect on the results.

Let's look at how using different bases in constructing rates can affect the results. In the USA data file, variable 125) FS$/PER uses the amount spent on the food stamp program as the numerator and the number of recipients of food stamps as the denominator for each state—so it is the *average benefit per recipient*. The variable 126) FS$/CAP uses the same numerator but uses the population of the state as the base, or denominator—so it is the *average benefit cost per capita*. Let's see if this difference in bases has any effect on relationships with other variables.

➤ *Data File:* **USA**
➤ *Task:* **Scatterplot**
➤ *Dependent Variable:* **126) FS$/CAP**
➤ *Independent Variable:* **86) % POOR**
➤ *View:* **Reg. Line**

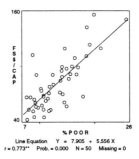

When we look at the scatterplot between the average benefit *per recipient* and the percent of people below the poverty line, we see that the correlation is .773 and that it is significant.

Now create a scatterplot between 125) FS$/PER (the average benefit cost *per capita*) and 86) % POOR.

Data File: **USA**
Task: **Scatterplot**
➤ *Dependent Variable:* **125) FS$/PER**
➤ *Independent Variable:* **86) % POOR**
➤ *View:* **Reg. Line**

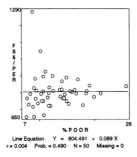

The correlation is .004 and not significant. Notice the case at the top of the screen. Removing this case from the scatterplot might have a sizable impact on the scatterplot.

We can use the outlier task to see if there is any single case that is making a substantial difference in the correlation coefficient. Click the [Outlier] option to see the effect of this outlier. The case at the top now has a red square around it. This case has been identified as the case that would have the greatest effect on the correlation coefficient if it were removed. We can see on the left side of the scatterplot that this case is Hawaii and that removal of this case would change the correlation coefficient from 0.004 to 0.165. The correlation would still not be significant. So there would be no reason to remove the outlier. Click on [Outlier] again to deselect this task.

Clearly in this example, one would reach quite different conclusions by using the cost per participant than by using the cost per capita. Poorer states spend more on food stamps per capita, probably reflecting that a greater percentage of the population receives food stamps. However, the poverty of the state does not appear to affect the amount received by the average recipient. We might speculate that the cost per recipient is a function of the food costs of the states. States in which food is more expensive will give greater benefits per recipient than will states in which food is less expensive. Let's see if we can check this idea.

We don't have a measure of the cost of food readily available, but we do have information on median rents. Generally, both rent and food reflect the relative cost of living. Let's look at the relationship between the food stamp cost per capita and the median rent within states.

Data File:	**USA**
Task:	**Scatterplot**
Dependent Variable:	**125) FS$/PER**
➤ *Independent Variable:*	**83) RENT**
➤ *View:*	**Reg. Line**

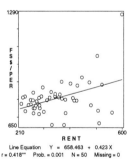

The correlation is .418 and significant. However, there is a case in the upper right corner—would removing this case have a significant impact on the correlation coefficient?

Click on the [Outlier] option to see the effect of the outlier. The outlier is Hawaii, and removal of this case would change the correlation from .418 to .235 and change the significance level from .001 to a value that is not significant. Click [Remove] to remove Hawaii. If you select the [Outlier] option again, you'll see that Alaska is the new outlier. Removal of Alaska would change the correlation to .122 and the relationship is not statistically significant. When Hawaii and Alaska are removed, there is no correlation between the two variables. This suggests that either food costs do not affect the size of food stamp benefits or median rent is not a good substitute for the cost of food.

The impact of Hawaii on this correlation coefficient demonstrates another frequent problem in comparative research. Because of the relatively small number of cases, a single case may have a sizable impact on the results. The wise comparative researcher will learn something about the units being used before beginning the actual analysis. In our example, if we knew that food costs for the 48 contiguous states had little variation, and food in Alaska and Hawaii was much more expensive, we would be better able to interpret the results. Similarly, if you are using a data set of nations, you should be aware that Singapore and Hong Kong are frequently included as cases in such data sets. However, these are "city-nations"—the boundaries of the city and the boundaries of the nation are the same. Since these cases have virtually no rural or farm areas, they are going to differ in many ways from other nations. Such deviant cases can have sizable effects on the results of an analysis.

With MicroCase, you can determine that an outlier is having a significant impact, but such a determination will not explain why the case is so deviant. In our example, we don't know why Hawaii is so different from the other cases and, consequently, we have no justification for removing Hawaii from the analysis.

If you have no idea why a particular case would be an outlier, you should double-check the data for that case. First, check the data values for the case against the original source. Perhaps the data were not entered correctly. Even if the data match the source, perhaps the source is incorrect—data are often misprinted or reported incorrectly. Try to find a second source for the data. If the two sources have decidedly different values, you must then determine which is the better source.

In some situations, you will know exactly why a case is an outlier. The results with that case removed may better reflect the overall relationship. You are then justified in removing that case from the analysis. Be sure to report any cases removed from analysis. In some situations, you may want to report the results both with and without the suspect case.

Removing a case from the analysis essentially assigns the missing data value to all variables for that case. This leads us to a third major difficulty in comparative research. Even without removing cases, comparative data sets frequently have a great deal of missing data—even worse, the cases with missing data almost never represent a random selection. For example, prior to the collapse of the Soviet bloc, data were much more difficult to find for Eastern European countries than for the other European countries. Similarly, data for less developed countries are more likely to be incomplete than are data for more developed countries. In the written exercises, you'll get a chance to see how missing data can influence the results.

By this point it should be pretty clear to you why rates, rather than raw numbers, are used in comparative research. In fact, this is a topic that was discussed back in Exercise 3—remember when you compared the *Playboy* circulation rates (circulation per 100,000 population) with the actual circulation rates for the 50 states? Let's take a quick look at this again.

Data File: **USA**
➤ *Task:* **Mapping**
➤ *Variable 1:* **58) PLAYBOY**
➤ *Variable 2:* **59) #PLAYBOY**
➤ *Views:* **Map**

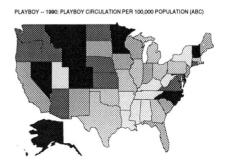

PLAYBOY -- 1990: PLAYBOY CIRCULATION PER 100,000 POPULATION (ABC)

r = –0.142

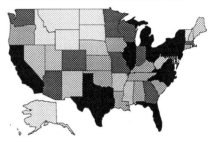

#PLAYBOY -- 1990: CIRCULATION OF PLAYBOY MAGAZINE (IN 1000S)

The top map shows the actual circulation of *Playboy* magazine for each state, whereas the bottom map shows the circulation per 100,000 population in each state. If we did not convert state-level variables to rates, nearly all variables we map would look the same because states with the largest populations will generally have more of everything—California has more hockey rinks than North Dakota simply because there are more people in California.

Most aggregate data are not published in the form of rates, so it is up to a researcher to turn variables into rates. Calculating a rate for each case in a data set would be extremely time-consuming if it had to be done manually. But most statistical analysis programs have a recode capability that allows you to compute rates electronically. The Windows 95 version of Student MicroCase has an especially powerful recode capability for creating rates. Let's see how it works. (If you are not using the Windows 95 version of Student MicroCase, skip ahead to the worksheet section.)

With the Windows 95 version of Student MicroCase, go to the FILE & DATA MENU and select the RECODE VARIABLES task. Then select the [Rate] option. The next screen prompts you to select the variable you want to turn into a rate. Select 59) #PLAYBOY as your numerator, but pause a moment to look at the variable description for #PLAYBOY. Notice that the coding for this variable is in 1,000s. That means that a state coded as 50 really represents 50,000 copies of the magazine that are sold (you need to move the decimal to the right three places). Hence, in order to properly calculate our rate, we need to indicate that the numerator is in 1,000s. So change the radio button that appears below the numerator box to "in 1,000s." Then click [Next] to continue.

You are now prompted to select your denominator. Since we want population as our denominator, select 2) POP 1990 as your denominator variable. Notice again that the description for the denominator indicates that the values are listed in thousands. So we again need to change the setting to "in 1,000s." Once you have done this, click [Next] to continue.

The next step is the trickiest because there are no steadfast rules. On this screen you have the ability to adjust the rate by multiplying both the numerator and denominator by a factor of 10. (Remember, if you multiply both the numerator and denominator by the same number, the ratio or rate remains the same.) You may wonder why it is necessary to adjust the rate at all. The reason for this is to make the results easier to interpret. For example, look at the first column in the grid at the bottom of the screen. This column, which appears in red, shows the results of your current recode: the number of *Playboy* magazines sold per person. Note, however, that these numbers are rather small and that there is not much difference between them—partially because of rounding. Thus, it would be better to convert these numbers into something larger so that we can better see the variation among the states.

As noted, the current multiplier for these results is 1 (which has no effect on the results). Click on the right "spinner" arrow once to increase the multiplier rate by 10. Note that this updates the rest of the information on the screen, including the first column of the grid at the bottom of the screen. Click the right multiplier button again to change the multiplier to 100. Since we have essentially moved the decimal two places to the left by multiplying by 100, the rate is also a percentage. These rates are still going to be hard to interpret because the percent of people who read *Playboy* ranges from 0.78 percent to 2.36 percent for all the states. So click the right multiplier arrow one more time to make the multiplier rate 1,000. Look again at the results in the first column of the grid. These values seem a little more useful. As you can see, there are approximately 9.4 *Playboy* magazines sold per 1,000 population in Alabama, whereas there are approximately 23.64 sold per 1,000 population in Alaska.

We could keep increasing the multiplier of our rate to 10,000 or 100,000 or more, but let's leave it as a circulation per thousand. Click [Next] to continue. You are now prompted to enter the name of the new variable. Type in the name **PLAYBOY/K** and modify the variable description to your liking. Click [Next] and the following screen indicates that the variable will be added to the end of your data file. Click [Finish] and then [OK] to return to the main menu.

Let's take a look at our handiwork.

Data File: **USA**
Task: **Mapping**
➤ *Variable 1:* **PLAYBOY/K**
➤ *View:* **Map**

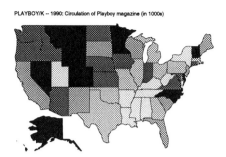

PLAYBOY/K -- 1990: Circulation of Playboy magazine (in 1000s)

It looks just like the map of 58) PLAYBOY we saw earlier, which is based on *Playboy* circulation per 100,000. (You should already know why the maps look identical, even though one is based on the circulation per 1,000 population and the other is based on the circulation per 100,000 population.)

Your turn.

Workbook exercises and software are copyrighted. Copying is prohibited by law.

1. You want to compare personal property taxes in different counties of a state. Which would be the better base to compute a rate: the total population of a county, the population over age 18, or the total number required to pay personal property taxes? (Circle one.)

   Total population

   Population over age 18

   Total number required to
   pay personal property taxes

   Explain your answer.

2. You want to determine the college graduation rate by state. Which would be the better base, the total population of the state or the population over age 25? (Circle one.)

   Total population

   Population over age 25

   Explain your answer.

3. You are interested in voter turnout by state in the last presidential election. Obviously, the numerator for the rate would be the number of persons in each state who voted in that election. List three possible bases, or denominators, that could be used for creating a rate.

    a. _____

    b. _____

    c. _____

    Circle the letter of the rate you think would be best, and explain why you think it would be the best.

4. Let's examine a variable from the USA data file.

    ➤ *Data File:* **USA**
    ➤ *Task:* **Mapping**

    a. What is the description for 63) %FEMALE LG?

    _____

    _____

    b. According to the description, what numerator was used in calculating this rate?

    _____

    c. According to the description, what base, or denominator, was used in calculating this rate?

    _____

d. Why wasn't the population of the state used as the base?

5. Traffic statistics use different bases. For example, traffic fatalities can be reported as number of fatalities per 1,000 population, number of fatalities per 1,000 licensed drivers, number of fatalities per 1,000 registered vehicles, and so on. In the USA data set, you have two measures of the amount of driving in each state: 127) MILES/DRV and 128) MILES/VHCL.

a. Look at the description of each of these variables and describe the difference in how they are calculated.

6. Let's examine several scatterplots in the USA data file.

a. Create the following scatterplot and answer the items that follow.

    Data File: **USA**
    ➤ Task: **Scatterplot**
    ➤ Dependent Variable: **70) PICKUPS**
    ➤ Independent Variable: **127) MILES/DRV**

    r = _____

    Prob. = _____

Click on [Outlier] to identify the case that has the greatest impact on the correlation coefficient. If this case were removed, what would the results be?

r = _____

Prob. = _____

Should you consider removing the outlier? (Circle one.)   Yes   No

b. Create the scatterplot below and answer the items that follow.

>
> *Data File:* **USA**
> *Task:* **Scatterplot**
> *Dependent Variable:* **70) PICKUPS**
> ➤ *Independent Variable:* **128) MILES/VHCL**

r = _____

Prob. = _____

Click on [Outlier] to identify the case that has the greatest impact on the correlation coefficient. If this case were removed, what would the results be?

r = _____

Prob. = _____

Should you consider removing the outlier? (Circle one.)   Yes   No

c. Why does the relationship with 70) PICKUPS differ depending on which measure is used? (Comparing the maps of these variables might help you answer this question.)

7. Create the scatterplot below and answer the items that follow.

> Data File: **USA**
> Task: **Scatterplot**
> ➤ Dependent Variable: **127) MILES/DRV**
> ➤ Independent Variable: **129) CARS/HSE90**

a. What is the description of 129) CARS/HSE90?

_____

_____

r = _____

Prob. = _____

Use the [Outlier] option to identify the case that has the greatest impact on the correlation coefficient. If this case were removed, what would the results be?

r = _____

Prob. = _____

Should you consider removing the outlier? (Circle one.)　　　Yes　No

b. Create the scatterplot below and answer the items that follow.

> Data File: **USA**
> Task: **Scatterplot**
> ➤ Dependent Variable: **128) MILES/VHCL**
> ➤ Independent Variable: **129) CARS/HSE90**

r = _____

Prob. = _____

Click on [Outlier] to identify the case that has the greatest impact on the correlation coefficient. If this case were removed, what would the results be?

r = _____

Prob. = _____

Should you consider removing the outlier? (Circle one.)　　　Yes　No

c. Why does the relationship with 129) CARS/HSE90 differ depending on which rate is used? (Comparing the maps of these variables might help you answer this question.)

8. Sometimes, researchers will include Washington, D.C., in a data set with the 50 states. Let's examine the data set named US–DC.

> *Data File:* **US–DC**
> *Task:* **Scatterplot**
> *Dependent Variable:* **3) %GRAD.DEG**
> *Independent Variable:* **2) %BLACK90**

a. Based on this scatterplot, fill in the following:

r = _____

Prob. = _____

b. Click on [Outlier] to identify the case that has the greatest impact on the correlation coefficient.

Name of outlier: _____

r = _____

Prob. = _____

c. Should you consider removing the outlier? (Circle one.)   Yes   No

d. What is your conclusion about the relationship between these two variables?

e. Create the scatterplot below and answer the items that follow.

> Data File: **US–DC**
> Task: **Scatterplot**
> ➤ Dependent Variable: **4) ONE P.HH90**
> ➤ Independent Variable: **3) %GRAD.DEG**

What is the description of 4) ONE P.HH90?

_____

_____

r = _____

Prob. = _____

f. Click on [Outlier] to identify the case that has the greatest impact on the correlation coefficient.

Name of outlier: _____

r = _____

Prob. = _____

Should you consider removing the outlier? (Circle one.)   Yes   No

g. What is your conclusion about the relationship between these two variables?

h. Would you advise researchers to include Washington, D.C., in a data set on the states of the United States? (Circle one.)   Yes   No

Explain your answer below.

9. The correlation function in the student version of MicroCase has been set to use listwise deletion of missing data. This means that, if a case has missing data on *any* of the variables included in the analysis, that case will be excluded from the entire list of correlations. Change back to the regular USA data file and create the following map:

> ➤ *Data File:*   **USA**
> ➤ *Task:*   **Mapping**
> ➤ *Primary Variable:* **51) % FAT**
> ➤ *View:*   **List: Rank**

a. Which case(s) has (have) missing data on this variable? (Scroll to the end of the list to see the missing data cases.)

> *Data File:*   **USA**
> *Task:*   **Mapping**
> ➤ *Primary Variable:* **80) FOODSTAMPS**
> ➤ *View:*   **List: Rank**

b. Which case(s) has (have) missing data on this variable?

> *Data File:*   **USA**
> *Task:*   **Mapping**
> ➤ *Primary Variable:* **92) MATH SCORE**
> ➤ *View:*   **List: Rank**

c. Which case(s) has (have) missing data on this variable?

10. Now let's look at correlations among these variables:

> Data File: **USA**
> ➤ Task: **Correlation**
> ➤ Select Variables: **51) % FAT**
> **80) FOODSTAMPS**

What is the correlation between these two variables? _____

How many cases (n) are used in the calculation?_____

Which cases are excluded?

> Data File: **USA**
> Task: **Correlation**
> ➤ Select Variables: **80) FOODSTAMPS**
> **92) MATH SCORE**

What is the correlation between these two variables? _____

How many cases are used in the calculation?_____

Which cases are excluded?

11. Now examine the correlations among all three variables at once and fill in the following correlation matrix.

	80) FOODSTAMPS	92) MATH SCORE	51) % FAT
80) FOODSTAMPS	_____	_____	_____
92) MATH SCORE	_____	_____	_____
51) % FAT	_____	_____	_____

a. How many cases are used in these calculations? _____

b. Do the results change when all three variables are in the matrix (compared with the results using only two variables)? If so, how?

c. Explain why researchers should be careful when working with data sets that contain considerable missing data.

12. Now, if you are using the Windows 95 version of MicroCase, let's do another recode. Variable 2) POP 1990 is population in thousands and 39) AREA is area in square miles. So, let's compute a population density figure by dividing population in thousands by area in square miles.

> Data File: **USA**
> ➤ Task: **Recode**
> ➤ Recode Option: **Rate**

For the numerator, select 2) POP 1990. Since our population variable is in 1,000s, click on the radio button that indicates "in 1,000s" and click [Next]. For the denominator, select 39) AREA and click [Next]. This yields a rate that shows the population in thousands per square mile. Convert this to population per square mile by clicking the right "spinner" button three times. Note that the number in the Multiplier column has changed to 1,000.

How many people per square mile are there in each of the following states?

Alabama	_____
Idaho	_____
New Jersey	_____

Click [Next] and provide a name for this new variable (you are limited to 10 characters). Click [Next], [Finish], and then [OK] to return to the main menu.

Data File: **USA**
➤ *Task:* **Mapping**
➤ *Primary Variable:* **Select the variable you created**
➤ *View:* **List: Rank**

Print out this table and attach it to this worksheet.

9

Field Methods

OVERVIEW

In this exercise, you will learn more about problems and issues in field research. Field research is less structured than other methods, and this has both advantages and disadvantages. You will see that field researchers must be careful not to let this lack of structure undermine their studies.

BEFORE YOU BEGIN

Please make sure you have read Chapter 9 in the textbook and can answer the following review questions (you need not write any answers):

1. When is field research better than other methods of social research?

2. Why is reliability often low in field studies, and how can it be increased?

3. Discuss the two major problems involved in entering the field.

4. What is the difference between structured and unstructured observation?

5. What are informants in field research? What are the advantages and disadvantages of using informants?

6. Describe field notes and the general processes by which they are analyzed.

7. What are the primary ethical concerns in field research?

Field research is considerably less structured than other types of social research. In survey research, we can look at how the sample was selected and at the questions asked. In comparative research, we can determine whether a particular rate appears to be an acceptable measure. But, in reports of field research, we have virtually no means of checking on the accuracy of the observations. This is why most field studies are exploratory, rather than hypothesis testing. In this exercise, we'll see some special problems that are created by this lack of structure.

One of the first problems is that different observers focus on different elements. For example, consider something as uncomplicated as a group of people talking with one another. One researcher might focus on the content of the conversation, while another might be more interested in nonverbal communication. Differences of style between males and females might be the main interest of another. Decoding the influence, or power structure, of the group might be another focus. Obviously, the interest of the researcher will influence what is observed and recorded, as well as what is not observed and not recorded. Of course, the purpose of field research is not only to observe, but also to organize and interpret these observations in a meaningful way.

Unfortunately, many researchers not only have certain interests, but also have developed definite opinions on a topic. For example, the researcher interested in gender styles of interaction already may believe that females will be submissive and males will be dominant. Bias—either intentional or unintentional—is almost certain to creep into this study. This researcher already has a hypothesis and is not using field research to explore gender styles in interaction but rather to document preconceived differences. This is not exploratory research but hypothesis-testing research. In this situation, other methods better suited to hypothesis testing would be more appropriate—content analysis of videotaped conversations might be one such approach.

At the other extreme, researchers who observe without any preconceived structure are bound to be buried in a sea of unrelated details and accomplish nothing. Thus the problem for the field researcher is to focus the study without predetermining its outcome.

Characteristics of the researcher also can have a major effect on the subjects of observation. An individual's actions are frequently influenced by who is watching. All of us engage in private behaviors that we would not want observed. We behave differently in front of our boss than we do in front of a subordinate. Age, gender, attractiveness, and many other characteristics of the observer can influence the behavior of those being observed. A 20-year-old woman and a 50-year-old man observing the same group might reach quite different conclusions since the behavior of the group may be affected by who is watching. In Japan, even the spoken language differs greatly according to the gender of the speaker; that is, men and women have different accents. This led to considerable embarrassment for some World War II American GIs who learned the language from their Japanese girlfriends. This is another reason that having more than one observer is a good idea.

Obviously, the effects of observer characteristics will be even more pronounced when informers are involved. In some societies, age, gender, and marital status determine interaction patterns—some members may not even be permitted to talk with the observer.

Field research also may be applied research. A researcher might observe the operation of a ward in a hospital. Sometimes, such studies are unstructured. For example, the researcher may discover that doctor/nurse conflicts are having a negative impact on patient care. More often, they are studies to *evaluate* an organization or process. The researchers already have a model about what "should" happen. The observations are used to determine the extent to which this "model" is being met. For example, business consultants frequently observe the operation of a company for a period of time in order to recommend changes that might improve the operation of the business. Typically, such researchers already know how "good" companies work—for example, that employees are sharing in the decision-making. Their observations are used to determine how the company should be changed to match this preconceived ideal. Such studies may be very useful for the organizations involved, but they are not social science.

NAME:

COURSE:

DATE:

EXERCISE

9

Workbook exercises and software are copyrighted. Copying is prohibited by law.

WORKSHEET

1. You want to study the ways in which a local public official tries to win re-election. There are at least two fundamentally different approaches to gaining access. Describe two approaches and the advantages and disadvantages of each approach.

   a. Describe the first approach.

   What are the advantages of the first approach?

   What are the disadvantages of the first approach?

   b. Describe the second approach.

   What are the advantages of the second approach?

   What are the disadvantages of the second approach?

2. Look at this photograph of new recruits to the Marine Corps to answer these questions:

a. How would a field researcher obtain access to this setting?

b. Boot camp is eight weeks long. What problems does this pose for the researcher?

c. What characteristics of the researcher might affect his or her success in completing a study in this setting?

d. Discuss the relative value of a new recruit versus that of an instructor as an informer.

e. List four different topics that a field researcher might study in Marine basic training.

1.

2.

3.

4.

f. What problems exist for taking field notes in a study of basic training?

g. If the recruits were female, how would this affect the research project?

3.   A field researcher who is observing police behavior remarks that she is sure the police officers are acting normally around her because they frequently use foul language and they often criticize their supervisors. Do you agree with her conclusion? Why or why not?

4.   a. A team of researchers is interested in studying the role of house-husband/stay-at-home dad. They are especially interested in how this role differs from that of the housewife/stay-at-home mother. How would you design a field study to explore this topic?

     b. List three problems the field researchers are likely to encounter and how you would solve them.

        1.

        2.

3.

5. Consider the following potential "informants" in the studies given. Discuss the pluses and minuses of trusting the information of each informant listed.

a. **In a study of police behavior in a small city:**

The mayor of the city

The chief of police in this city

The prosecutor in this city

A defense lawyer

A prisoner in the city jail

b. **In a study of a school for rebellious teenagers:**

The father of a recent graduate who appears to have become better adjusted

A student who was dismissed for incorrigible behavior

A long-time instructor at the school

The director of a similar, competing school

The full-time maintenance man who lives on the school campus

The counselor who meets with each student once a week to assess progress

6. A researcher pretends to be a supporter in order to infiltrate an antigovernment organization that sometimes commits illegal acts. What ethical problems are most relevant here?

7. A social scientist who is a member of an organization decides to do a field study of this organization and receives funding from the organization for the study. Describe the problems in this situation.

8. A social scientist who has very strong views on the abortion controversy decides to do a field study of an organization that works for the other side on this issue. Describe the problems in this situation.

Experimental Methods

OVERVIEW

In this exercise, you will learn more about the necessary conditions for an experiment and about the power of the experiment in eliminating sources of spuriousness in relationships. You also will gain experience in designing experimental research and drawing the proper conclusions from the results.

BEFORE YOU BEGIN

Please make sure you have read Chapter 10 in the textbook and can answer the following review questions (you need not write any answers):

1. What are the two essential characteristics of true experiments?

2. Compare the advantages and disadvantages of laboratory experiments and field experiments.

3. What is the difference between an experimental group and a control group?

4. What is a double-blind experiment, and what use does it serve?

5. How can experiments be used in surveys?

6. What two primary sources of bias in experiments can reduce reliability?

7. Why is it important to make sure in an experiment that the independent variable really varied in the way it was supposed to vary?

8. Describe how each of the following factors can affect the results of quasi-experiments: history, maturation, testing effects, instrument effects, regression to the mean, and selection bias.

9. What is subject "mortality," and how can it affect experimental results?

10. What are the primary ethical concerns in experimental research?

The social scientists who designed the questionnaire for the 1996 General Social Survey were interested in how question wordings on government spending might affect the responses—would relatively minor changes in the phrasing of questions have a significant impact on the distribution of responses? They recognized that an experiment would be the best method for testing for such an effect. Half of the survey respondents, randomly selected, would be given one question wording, and the other half would be given the second question wording. The responses of the two groups could then be compared. The independent variable would be the question wording—there are two different treatment conditions represented by the two different wordings. The dependent variable would be the response to the question. By comparing the responses of subjects in the two treatment conditions, researchers could tell whether question wording had an effect.

Let's analyze the results of this experiment. Because the GSS is a survey, the GSS data file is organized in a manner appropriate for survey analysis. Consequently, we have created a data set named EXPER that is better organized for experimental analysis.

> *Data File:* **EXPER**
> > *Task:* **Cross-tabulation**

Let's examine the variable list for this file. The value of the first variable is the experimental condition for that subject—were the respondents asked the first wording of the question or the second wording? The second variable is the response to the first question. The first treatment used the following wording: Are we spending too much, too little, or about the right amount on *welfare*? The second treatment had different wording: Are we spending too much, too little, or about the right amount on *assistance to the poor*? We want to compare the responses of those in treatment 1 with the responses of those in treatment 2. You can see that there are a series of additional questions about spending in which the wording differed slightly between the two treatment conditions. So, in fact, we have a series of experiments.

Because the cases were *randomly assigned* to conditions, other characteristics of the cases will not be correlated with the independent variable (i.e., the question wording). We can see this by looking at the cross-tabulation of the treatment variable with these other characteristics.

	Data File:	**EXPER**
	Task:	**Cross-tabulation**
➤	*Row Variable:*	**13) SEX**
➤	*Column Variable:*	**1) TREATMENT**
	➤ *View:*	**Tables**
	➤ *Display:*	**Column %**

1) TREATMENT		
	One	Two
13) MALE	44.2%	44.3%
SEX FEMALE	55.8%	55.7%
TOTAL	100.0%	100.0%

V=0.001

We can see that the distribution of sex is almost identical in the two conditions—44.2 percent of the subjects in the first condition and 44.3 percent of the subjects in the second condition were male. Other characteristics of the subjects would also be about the same across the two conditions. The fact that the independent variable is uncorrelated with other characteristics of the subjects is the strength of experimental research. Remember, for a third variable to make a relationship spurious, that variable must be correlated with both the independent and dependent variables. So in experiments we do not have to worry about possible sources of spuriousness. We only need to examine the relationship between the independent and dependent variables to support or reject our research hypothesis.

Data File: **EXPER**
Task: **Cross-tabulation**
➤ *Row Variable:* **2) WELFARE**
➤ *Column Variable:* **1) TREATMENT**
➤ *View:* **Tables**
➤ *Display:* **Column %**

| | 1) TREATMENT | |
	One	Two
2) Too little	15.6%	56.2%
W Right	26.8%	25.8%
E Too much	57.7%	18.0%
L		
F		
A		
R		
E		
TOTAL	100.0%	100.0%

V=0.468**

We can see that 15.6 percent of those shown the first wording (i.e., *spending on welfare*) thought too little was being spent, while 56.2 percent of those in the second treatment condition (i.e., *assistance to the poor*) felt this way. Let's look at the statistical summary.

Data File: **EXPER**
Task: **Cross-tabulation**
Row Variable: **2) WELFARE**
Column Variable: **1) TREATMENT**
➤ *View:* **Statistics (Summary)**

WELFARE   by   TREATMENT

Nominal Statistics

Chi-Square: 805.148 (DF = 2; Prob. = 0.000)
V: 0.468   C: 0.424
Lambda: 0.403   Lambda: 0.305   Lambda: 0.349
(DV=1)   (DV=2)

Ordinal Statistics

Gamma: -0.685   Tau-b: -0.442   Tau-c: -0.507
s.error   0.024   s.error   0.015   s.error   0.017

Dyx: -0.507   Dxy: -0.385
s.error   0.017   s.error   0.013

Prob. =   0.000

We can see that the results are statistically significant. So the difference in wording did have an effect on the subject's view toward government spending.

If we can assume that the dependent variable is measured at the interval or ratio level, there is another technique that can be used in experimental analysis. Analysis of variance is similar to regression analysis except the independent variable is nominal, or categorical. To demonstrate this technique, we'll assume that the dependent variables are close enough to interval to justify its use.

Data File: **EXPER**
➤ Task: **ANOVA**
➤ Dependent Variable: **2) WELFARE**
➤ Independent Variable: **1) TREATMENT**
➤ View: **Graph**

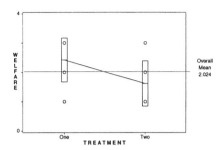

This is called a *box-and-whisker diagram*. It is just a scatterplot between the two variables with some additional information. The independent variable, treatment condition, is shown across the bottom. Those in the first treatment condition are shown at the left, and those in the second treatment condition are shown at the right. The vertical axis represents the dependent variable. In this example, there are only three possible values for this variable, so there are many cases at each of the three levels but only one dot is visible. So far, this is just the same as a scatterplot.

Since the independent variable in analysis of variance is categorical, we can provide additional information about the dependent variable within each of these categories. We can calculate the mean and variation of the dependent variable within each category of the independent variable. The vertical box shown at the left represents the variation in the dependent variable for the first condition. The horizontal line (the whisker) in the middle of this box represents the mean of the dependent variable. The box at the right shows the same information for the second condition. A line connects the means of the two conditions. In this example, there is a sizable difference in the means.

Let's look at the analysis of variance (ANOVA) results.

Data File: **EXPER**
Task: **ANOVA**
Dependent Variable: **2) WELFARE**
Independent Variable: **1) TREATMENT**
➤ View: **ANOVA**

Analysis Of Variance
Dependent Variable: WELFARE
Independent Variable: TREATMENT   N: 2760   Missing: 144

ETA Square = 0.219

TEST FOR NON-LINEARITY:
Not appropriate when only 2 categories for Independent Variable.

Source	Sum of Squares	DF	Mean Square	F	Prob.
Between	445.125	1	445.125	772.915	0.000
Within	1588.344	2758	0.576		
TOTAL	2033.469	2759			

This is similar to the analysis of variance table shown for regression. For our needs, we can ignore all the information except the significance level (Prob. = 0.000) presented next to the F value. The eta-squared value (0.219) is a measure of association and tells us the strength of the relationship between the two variables.[1] In this example, this relationship is highly significant. Now select the [Means] option.

Data File: **EXPER**
Task: **ANOVA**
Dependent Variable: **2) WELFARE**
Independent Variable: **1) TREATMENT**
➤ *View:* **Means**

Means, Standard Deviations and Number of Cases of Dependent Var:
WELFARE
by Categories of Independent Var: TREATMENT
Difference of means across groups is statistically significant (prob. 0.000)

	N	Mean	Std.Dev.
One	1394	2.421	0.745
Two	1366	1.618	0.772

This is the mean of the dependent variable for each condition. For the 1,394 cases in the first treatment condition, the mean was 2.421—just slightly under the midpoint for the categories "right" and "too much." For the 1,366 cases in the second condition, the mean was 1.618—or just slightly above the midpoint for the categories "too little" and "right."

Let's now look at the wording of two questions relating to spending on the space program. Again, half the respondents were asked one version of the question (government spending on the *space exploration program*), while the second half were asked another version (government spending on *space exploration*).

Data File: **EXPER**
Task: **ANOVA**
➤ *Dependent Variable:* **3) SPACE PRG**
➤ *Independent Variable:* **1) TREATMENT**
➤ *View:* **Graph**

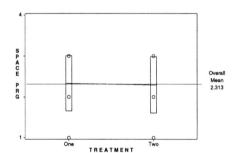

In this example, the line that connects the means is horizontal. There is virtually no difference between the conditions. If you look at the analysis of variance table, you see that the results are not significant and eta-squared is zero.

Here is another set of questions relating to government spending on health. (Examine the variable description for 4) HEALTH to see the difference in question wording.)

---

[1] Eta and r, the Pearson correlation coefficient, are based on the same mathematical ideas except that r assumes that the relationship will be linear and eta has no restrictions on the nature of the relationship.

Data File: **EXPER**
Task: **ANOVA**
➤ *Dependent Variable:* **4) HEALTH**
➤ *Independent Variable:* **1) TREATMENT**
➤ *View:* **Graph**

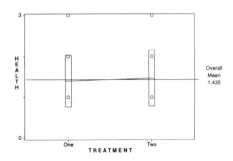

Notice that the line connecting means is almost horizontal. Let's look at the actual means.

Data File: **EXPER**
Task: **ANOVA**
Dependent Variable: **4) HEALTH**
Independent Variable: **1) TREATMENT**
➤ *View:* **Means**

Means, Standard Deviations and Number of Cases of Dependent Var: HEALTH

by Categories of Independent Var: TREATMENT
Difference of means across groups is statistically significant (prob. 0.005)

	N	Mean	Std.Dev.
One	1394	1.400	0.634
Two	1372	1.470	0.686

Treatment 1 with 1,394 cases has a mean of 1.400, while treatment 2 with 1,372 cases has a mean of 1.470. Not much difference. Now let's look at the analysis of variance table.

Data File: **EXPER**
Task: **ANOVA**
Dependent Variable: **4) HEALTH**
Independent Variable: **1) TREATMENT**
➤ *View:* **ANOVA**

Analysis Of Variance
Dependent Variable: HEALTH
Independent Variable: TREATMENT   N: 2766   Missing: 138

ETA Square = 0.003

TEST FOR NON-LINEARITY:
Not appropriate when only 2 categories for Independent Variable.

Source	Sum of Squares	DF	Mean Square	F	Prob.
Between	3.441	1	3.441	7.886	0.005
Within	1206.214	2764	0.436		
TOTAL	1209.656	2765			

Eta-squared is .003 and is significant at the .005 level. How can such a weak relationship be significant?

Remember that statistical significance tells us only that it is likely that a relationship exists in the population. Two elements affect this level: the strength of the relationship in the sample and the size of the sample. In this study, the sample size is extremely large so that even a small relationship is likely to be significant. Again this demonstrates why we should look at both the significance level and the strength of the relationship. The eta-squared of .003 tells us that, while a relationship may exist in the population, it is a very weak relationship.

We can see that the analysis of experiments is much simpler than the analysis of surveys because we need not worry about sources of spuriousness.

What do these results tell us? Simply that relatively minor changes in wording can have a significant effect on responses. (We examined this issue in a slightly different way in Exercise 3.) Suppose that instead of doing an experiment, researchers had used one set of question wordings in the GSS one year and the other set of wordings the next year. If we observed a difference in the responses between the two years, we could not be sure that it was caused by the change in question wordings. It is possible that the attitudes in the population actually changed during the year. By using an experiment, we have eliminated any such alternative explanations and can be sure that the question wordings themselves are causing the differences.

This experiment is unusual in that a probability sample of a particular population provided the subjects for the experiment. This means that the results of this study can be generalized to the relevant population. However, this isn't really very important because our interest wasn't in this specific population but in the more general issue of the importance of question wordings. The use of a probability sample also means that it is possible to examine the effect of other variables, such as education, income, and race, on attitudes toward spending. Such analysis, of course, would be regular survey analysis and you would have to consider possible sources of spuriousness.

Your turn.

NAME:

COURSE:

DATE:

EXERCISE

10

Workbook exercises and software are copyrighted. Copying is prohibited by law.

WORKSHEET

1. The text describes an experiment in which students are randomly assigned to an experimental group that, over a period of six months, watches a series of movies designed to reduce racial prejudice or to a control group that does not watch the movies.

    a. Suppose a large number of subjects dropped out of the experimental condition—they stopped staying after school to watch the movies. What problem would this introduce in reaching a conclusion about the effect of the movies?

    b. How could the design of the experiment be changed to minimize this problem?

    c. Suppose that on the posttest there was no difference in racial prejudice between the two groups—individuals in both groups were extremely low in prejudice. In addition, there was no change between pretest and posttest for individuals in either group. Would this indicate that the movies had no effect on prejudice? Why or why not?

d. How could this study be modified to provide a better test of the effectiveness of the movies?

e. Suppose that on the posttest there was no difference in racial prejudice between the two groups—individuals in both groups were low in prejudice. However, in both groups, there was a big decrease in racial prejudice between the pretest and the posttest for each individual. Analyzing the friendship structure and interpersonal contact among all subjects, the researchers find that subjects in the control group had many friends and a great deal of interpersonal contact with subjects in the experimental group. How might this fact have affected the results of the study?

f. How could this study be modified to eliminate this problem?

2. Two social scientists were interested in studying the effect of sex education on sexual behavior. They selected a school in which some of the students would be participating in a sex education study during the year while others would be taking a computer literacy course. (Parents selected one or the other of these courses for their children.) The researchers administered a questionnaire on sexual behavior at the end of the year and found that students who had taken the sex education sequence were much more likely to be sexually active than were the other students. They concluded that sex education increases the sexual activity of preteens.

a. Given the design of this study, can we have any confidence in the conclusion? (Circle one.)   Yes   No

When the researchers were told that this was not an experiment, they responded, "That's OK. It's a quasi-experiment."

b. Describe why this study is not a true experiment.

c. Provide another explanation for why the researchers found the two groups to differ in sexual activity.

d. Assume that the researchers' hypothesis was that taking a sex education class would increase the sexual activity of the participants. How could they have designed a true experiment to test this hypothesis?

e. What ethical problems would the researchers encounter in this study, and how could they solve them?

3. Does taking a college course on American politics increase the political knowledge of students? Let's examine two approaches to answering this research question.

   a. At the end of the first term of the school year, a researcher measures political knowledge among two samples of first-year students: a sample of those who took the American politics course and a sample of those who did not take the course. Students who took the course have substantially higher political knowledge than those who did not take the course. The researcher concluded that taking the course on American politics does increase the political knowledge of students. What flaws are there in this research design?

   b. A second researcher uses a different approach. At the beginning of the term, the researcher measures political knowledge among all students who are taking the American politics course. At the end of the course, the researcher again measures political knowledge among all students who completed the course. At the end of the course, students who completed the course have much higher political knowledge than they did at the beginning of the course. The researcher concludes that taking the American politics course does increase the political knowledge of students. What flaws are there in this research design?

4. Now let's use the EXPER file to fill in the column percentages in the table below:

> ➤ *Data File:* **EXPER**
> ➤ *Task:* **Cross-tabulation**
> ➤ *Row Variable:* **14) RACE**
> ➤ *Column Variable:* **1) TREATMENT**
> ➤ *View:* **Tables**
> ➤ *Display:* **Column %**

	One	Two
White	_____%	_____%
Black	_____%	_____%
Other	_____%	_____%

V = _____

Prob. = _____

Why is the racial distribution so similar across the two treatment conditions?

5. Variables 5 to 12 were also part of the methodological experiment. Let's return to the ANOVA task.

   a. Select one variable from among variables 5 through 12 as the dependent variable, and use 1) TREATMENT as the independent variable.

   Which variable did you select?   _____

   What is the description of this variable?

   _____

   _____

   What is the mean of each group?

   Treatment 1: mean is _____

   Treatment 2: mean is _____

   Is this difference statistically significant? (Circle one.)   Yes   No

b. Select a second variable from among variables 5 through 12, and conduct
   the exact same analysis.

   Which variable did you select?   _____

   What is the description of this variable?

   _____

   _____

   What is the mean of each group?

   Treatment 1: mean is _____

   Treatment 2: mean is _____

   Is this difference statistically significant? (Circle one.)   Yes   No

6. Is an argument for some position more persuasive if it is presented by a male?
   Assuming that you had a large group of volunteers for an experiment, design
   an experiment to test the proposition that people will be more persuaded by
   an argument if it is presented by a male. Present a diagram—roughly following Design 4 in the textbook—along with your description of the research
   design. (Note: Keep in mind that an experiment might have two or more
   experimental groups rather than one experimental group and one control
   group.)

7. In the first class session for a course, the professor announces that each person in the class has been randomly assigned to one of two groups, A or B. Group A will use the regular textbook for the course, but group B will use a computerized tutorial—a tutorial that the professor has developed and believes will greatly increase student learning while also making it much more enjoyable. The two groups will otherwise receive the same treatment (same lectures, exams, etc.). The professor announces that the purpose of this experiment is to determine whether the computerized tutorial increases learning. At the end of the course, the professor will compare the exam scores of the two groups to determine whether the computerized tutorial was more effective than the textbook.

a. Describe how subject bias (demand characteristics) could be a problem here.

b. Describe how experimenter bias could be a problem here.

c. What ethical issue might be raised by some of the students in the course? How might this problem be resolved?

Content Analysis

OVERVIEW

In this exercise, you will learn more about content analysis by doing a small content analysis of newspaper personal ads in which people are trying to meet people of the opposite sex. You will code some of the variables and then test several hypotheses based on the data file.

BEFORE YOU BEGIN

Please make sure you have read Chapter 11 in the textbook and can answer the following review questions (you need not write any answers):

1. What is content analysis, and what kinds of materials can be used in such analysis?

2. What is the difference between manifest content and latent content?

3. Describe the two primary factors that determine reliability in content analysis.

4. How do deposit bias and survival bias affect the validity of content analysis?

5. What does it mean to say that content analysis is a form of unobtrusive measurement?

The unit of analysis in content analysis is not an individual, nor even a collection of individuals, but rather a cultural artifact. However, the basic steps of the research process remain the same. After identifying the topic of interest and developing the research hypotheses, the researcher selects an appropriate sample of cases (artifacts), develops measures of the concepts (the coding system), collects the data (codes the cases), and tests the hypotheses.

In this exercise, you will apply content analysis to personal ads in which people are seeking to meet persons of the opposite sex. These ads were selected from one issue of a daily newspaper in the Pacific Northwest. In this particular newspaper, the newspaper ads are free. Each advertiser records a voice greeting that plays when an interested person calls the published number. Those who respond to the ads pay $1.99 per minute to hear the message at the 1-900 number and to leave a message for the advertiser.

Since there were an equal number of male and female ads, a stratified random sampling design was used. The ads were first divided into those placed by men and those placed by women. Then 50 ads were selected randomly from each gender group. The ads included in the sample are reprinted at the end of this discussion—the first 50 are written by males and the second 50 are written by females.

Most of the ads provide information about the ad writer and describe characteristics of the person they are seeking. The content of these ads can be coded easily. In fact, if you look at the top of page , you'll see that some elements already have been coded by the advertiser. For example, W means "white," C means "Christian," LTR means "long-term relationship," and so on. Looking over the ads, we decided upon the following variables and categories:

GENDER
Gender of advertiser:
        1—male, 2—female

AGE-AD
Age of advertiser:
        1—under 40, 2—40 & over, 3—not stated
        If range given, code youngest end.

AGE-DES
Age desired:
        1—under 40, 2—40 & over, 3—not stated
        If range given, code youngest end.

LTR
Long-term relationship (LTR):
        1—LTR stated, 2—not clear, 3—no long term
        If indicate "friends first," code as 2.

SEXUAL IMP
Sexual implication:
        1—explicit, 2—implicit, 3—none
        If indicate "friends first," code as 3.

APPEAR-AD
Physical description of advertiser:
> 1—none provided, 2—minimal description provided (e.g., height/ weight proportionate or height), 3—stresses attractiveness (e.g., cute, beautiful, handsome, good looking)

APPEAR-DES
Physical description desired:
> 1—none provided, 2—minimal description provided (e.g., height/ weight proportionate or height), 3—stresses attractiveness (e.g., cute, beautiful, handsome, good looking)

SECUR
Financial security desired:
> 1—stated, 2—none indicated
> If seeking professional, code as 1.

ACTIVE
Stresses active lifestyle:
> 1—yes, 2—no

The data file for this project is named CONTENT. Open this file and select the LIST DATA task.

> ➤ *Data File:* **CONTENT**
> > ➤ *Task:* **List Data**

In the Windows 95 version of Student MicroCase, go to the FILE & DATA MENU and select LIST DATA. Then click [OK] to accept the settings on the pop-up window.

The program will now list the data for all variables and all cases. Each case is one of the ads. The order of the cases matches the order in the list of ads. The variables have been defined and data have been entered for most of the variables.

Let's look at the coding of the first case:

1. CHRISTIAN AND KIND
SWM, 23, Christian values, 6'1", physically fit, attractive, enjoys summer activities, beaches, conversation. Would like to meet a Christian female for friendship, possibly more.

We now can examine the codes for each variable. (Use the scroll bar to move to the variables not currently showing.) This case has been coded 1 for male for the first variable (GENDER); variable 2 (AGE-AD) is coded as 1, which represents the under 40 category, because the age of the advertiser is indicated as 23; the desired age is not stated for the first case, so variable 3 (AGE-DES) is coded as 3. Notice how this coding is matching the row for the first case in the data set. The ad states that the advertiser is looking for "friendship, possibly more." This is coded as a 2 according to the "friendship first" note for variable 4. There are no sexual implications in the ad so variable 5, SEXUAL IMP, is coded 3. The advertiser describes himself as "attractive"—this places him in category 3 of variable 6,

APPEAR-AD. He mentions neither the appearance nor the financial security of the person he is seeking, so variable 7, APPEAR-DES, is coded as 1 and variable 8, SECUR, is coded as 2. Finally, he is coded as 1 on variable 9, ACTIVE. If you were to list the data for all cases, you would see that each case has already been coded for these first nine variables.

We're going to test the hypothesis that the sex of the advertiser affects the content of the ads. But first, we need to determine if our coding is reliable enough for analysis. In a previous exercise, you learned about the use of rekey verification to increase the accuracy of data entry from questionnaires. Rekey verification is useful for checking data entry mistakes—the correct value is obvious. In content analysis, the "correct" answer is subject to interpretation. The researcher wants to be able to compare the data from multiple coders and assess the degree of concensus. Data from each of the coders must be available to the researcher, so rekey verification is not an appropriate technique. Instead, each additional coder should enter data into new variables.

Note that no data have been entered for variables 10 through 18. These variables are essentially the same as variables 1–9, except that the variable names include an extra character. These variables will be used by a second coder (that being you) to reenter the same classified advertising data. In this way you will be able to compare your coding to that determined (and entered) by the original coder. Again, by comparing the coding of the same data by two different coders, we can assess the reliability and accuracy of our data. In the exercises, you'll be asked to code several cases on some of these variables, so let's quickly see how to enter the data using MicroCase.

Exit the LIST DATA task and return to the main menu. With the CONTENT data file open, select the ENTER DATA task.

> *Data File:* **CONTENT**
> ➤ *Task:* **Enter Data**

In the Windows 95 version of Student MicroCase, a pop-up window will appear when you select the ENTER DATA task. Click [OK] to accept the default settings as Grid Entry format. Then click on the cell in the upper left corner to position the cursor for data entry.

Let's enter the codes (i.e., the numeric categories) for the first two cases for variables 10 through 18. We just discussed the codes for the first case, so we'll reenter those codes for variables 10 through 18. Here is a summary of how the first case was coded on each of the nine variables:

	Code	Category Label
GENDER1	1	male
AGE-AD1	1	under 40
AGE-DES1	3	not stated

LTR1	2	not clear
SEXUAL IM1	3	none
APPEAR-AD1	3	stresses attractiveness
APPEAR-DE1	1	none provided
SECUR1	2	none indicated
ACTIVE1	1	yes

After you've entered the code for variable 18, the program will go to variable 10 for case 2. The first case has a value of 1 for the variable 10, GENDER1. To enter this value, simply type **1** and press the <Enter> key. The cursor will move to the next variable, AGE-AD1, for the first case. Type **1** and press the <Enter> key. Enter values for the remaining variables in the same manner. If you make a mistake, just click on the appropriate cell and reenter the value. If you accidentally enter a value which is out of the appropriate range, the program will refuse to accept the value. For example, if you try to use the value 3 for the gender of the individual, a message will tell you that this value is out of range—you will then want to enter the correct value. Enter the remaining data for the first case.

Let's take a look at the second case and code it.

2. CHRISTIAN IN SEATTLE
Attractive, Christian SWM, 28, 6', H/W proportionate, active person, financially secure, stable and healthy. Would like to meet Christian female, for friendship, possibly more.

This is very similar to the first ad. Refer back to the original coding scheme and fill in the codes for this case.

	<u>Code</u>
GENDER1	_____
AGE-AD1	_____
AGE-DES1	_____
LTR1	_____
SEXUAL IM1	_____
APPEAR-AD1	_____
APPEAR-DE1	_____
SECUR1	_____
ACTIVE1	_____

Let's see if your coding is accurate. You should have coded GENDER1 as 1; for AGE-AD1 the coding is also 1; AGE-DES1 is 3. Again the phrase "friendship, possibly more" leads us to code 2 for LTR1 and 3 for SEXUAL IM1. The advertiser provides a minimal physical description of himself (6', H/W proportionate) but does not stress attractiveness—APPEAR-AD1 should be coded as 2. He neither provides a physical description of the person sought (APPEAR-DE1 should be coded as 1) nor requests financial security (SECUR1 is coded 2). He does indicate an active lifestyle—code 1 for ACTIVE1. You now have completed coding for the first two cases. Return to the main menu and answer Yes to the prompt about saving changes to the data file.

When coders read and code a large number of cases, they unintentionally may begin to modify the meaning of the codes. This is known as coder drift—criteria used for the later cases drift from those used for the earlier cases. The more subjective the coding, the more likely this problem is to occur. In this example, coder drift is much more of a problem when coding sexual implications of the ad than when coding the gender of the advertiser. This is another reason for having detailed coding instructions that the coder often refers back to and for using multiple coders.

For the present, let's assume that the codings are reliable and do a preliminary test of our hypothesis that the content of the ads will differ by gender. But before we select an analysis task, let's have the category "not stated" temporarily changed to missing data. We've used this feature several times before, so it should be familiar to you by now. If you are using the Windows 95 version of the software, select the FILE SETTINGS option from the main menu (if you are not using the Windows 95 version of the software, simply click on the large button [Set Categories to Missing Data]). In one of the blank boxes, type the phrase **Not Stated**, and then click [OK]. In all subsequent analysis, cases coded NOT STATED on a variable will be assigned the missing data value on that variable.

Now let's see whether there is a difference between males and females in terms of whether they stated age requirements in their advertisements. We'll use the Cross-tabulation task to analyze this.

Data File:	**CONTENT**
➤ Task:	**Cross-tabulation**
➤ Row Variable:	**3) AGE-DES**
➤ Column Variable:	**1) GENDER**
➤ View:	**Tables**
➤ Display:	**Column %**

1) GENDER	MALE	FEMALE
3) UNDER 40	90.9%	40.0%
AGE-DES 40 & OVER	9.1%	60.0%
TOTAL	100.0%	100.0%

V=0.545**

We can see that males are more likely to seek someone under 40 than are females. Remember, however, that we must consider the possibility of spurious relationships. Perhaps there is some other factor creating this relationship. An obvious possibility might be the age of the advertiser—perhaps the males are younger than the females, and when we control for the advertiser's age, there may be no difference. Let's try this analysis.

Control: Under 40

Data File:	**CONTENT**
Task:	**Cross-tabulation**
Row Variable:	**3) AGE-DES**
Column Variable:	**1) GENDER**
➤ Control Variable:	**2) AGE-AD**
➤ View:	**Tables (Under 40)**
➤ Display:	**Column %**

	1) GENDER	
	MALE	FEMALE
3) UNDER 40	100.0%	87.5%
AGE-DES 40 & OVER	0.0%	12.5%
TOTAL	100.0%	100.0%

V=0.304

Just a reminder, the option for selecting a control variable is located on the same screen you use to select the other variables. For this example, select 2) AGE-AD as a control variable and then click [OK] to continue. Separate tables for each of the AGE-AD categories will be shown for the AGE-DES and GENDER cross-tabulation. The first table includes only those advertisers who are under 40 years of age.

This first table includes only those cases who are under 40 years of age. Examine the differences between gender. Let's look at the next control table: those advertisers who are over the age of 40.

Control: 40 & Over

Data File:	**CONTENT**
Task:	**Cross-tabulation**
Row Variable:	**3) AGE-DES**
Column Variable:	**1) GENDER**
➤ Control Variable:	**2) AGE-AD**
➤ View:	**Tables (40 & Over)**
➤ Display:	**Column %**

	1) GENDER	
	MALE	FEMALE
3) UNDER 40	90.0%	20.0%
AGE-DES 40 & OVER	10.0%	80.0%
TOTAL	100.0%	100.0%

V=0.704**

Click the appropriate button at the bottom of the task bar to look at the second (or "next") partial table for AGE-AD.

Unfortunately, there are not enough cases to adequately test this alternative. The largest group, males under 40, has only 20 cases. The other groups have 10 or fewer. With only 100 cases, we are limited to examining bivariate relationships. If we were seriously interested in this issue, we would need to obtain a much larger sample of ads.

Your turn.

LIST OF ADVERTISERS

Use these cases and the code sheet that appears at the end of this list when doing Exercise 11.

Codes used by advertisers: LTR=Long-Term Relationship C=Christian J=Jew W=White B=Black H=Hispanic A=Asian N/S=Non-smoker N/D=Non-drinker S=Single D=Divorced M=Male F=Female H/W=Height/weight D/D-free=Drug- and disease-free P=Professional

1. CHRISTIAN AND KIND

SWM, 23, Christian values, 6'1", physically fit, attractive, enjoys summer activities, beaches, conversation. Would like to meet a Christian female for friendship, possibly more.

2. CHRISTIAN IN SEATTLE

Attractive, Christian SWM, 28, 6', H/W proportionate, active person, financially secure, stable and healthy. Would like to meet Christian female, for friendship, possibly more.

3. LOOKING FOR MS. RIGHT

SWM 23, 6'3", 165 lbs., brown/blue, enjoys camping, hiking, outdoors, movies, quiet nights. Seeking SF, 19–25, with similar interests. N/S, N/D.

4. A CUT ABOVE

Handsome, caring SWM, 47, 5'0", 190 lbs., honest, intelligent, business owner, enjoys arts, outdoors, travel. Seeking fit, classy lady, comfy in jeans/high heels, for LTR.

5. HARD-WORKING MAN

Handsome DWM, 40, 6'1", blond/blue, beard, country gentleman enjoys outdoors, camp fires, barbecues. Seeks attractive country girl, 35–45, N/S, light drinker for LTR.

6. BLACK GOLD

SBM, 37, 6'2", handsome, fit, romantic, easygoing, fun-loving, liberal, enjoys movies, sports, the arts, travel, dancing. Seeking lady with similar interests. 5'4"+ for LTR.

7. DON'T ANSWER THIS

Unless you seek a WM, 35, blonde/blue, who likes romantic nights and the outdoors. Please be WP/HF.

8. ROMANTIC TURK

Handsome, loving and passionate DWM, 46, 5'6", 140 lbs., willing to commit to a LTR, with active, fit, N/S lady 33–45, into personal and spiritual growth.

9. CARING KENT MAN

Thoughtful, caring SM, 39, 6', 185 lbs., brown/blue, enjoys motorcycles, music, movies, etc. Seeking SF, 21–38, H/W proportionate with similar interests, for possible LTR.

10. CHEF IN THE HOUSE

Attractive, Christian SBM, 30, N/S, N/D, easygoing, nice person, seeks very romantic, caring and honest SF friend, 25–33, N/S, N/D.

11. I COULD BE YOUR MAN

Are you 29–44, professional H/WF and have Christian values? Professional, athletic DWM, 42, hazel eyes, 185 lbs., seeks woman to pamper and adore.

12. HIGH ENERGY WOMAN

Work/play hard? Me too. SWPM, 38, 5'10", slim, handsome. Not a homebody. Love R&B music, dancing, biking, talking, laughing, cuddling. Seeking SWPF 30–40.

13. TRUE MAN FOR FRIENDSHIP

Sensitive, intelligent, honest, handsome SWM, 40, 5'10", 160 lbs. Seeking partner to share life's pleasures. Attractive, proportionate, adventurous WF, 35–45. Emotionally available, monogamous LTR.

14. RENAISSANCE MAN

Handsome, Eastside WD/PCM, 45, business owner, 5'10", handsome, brown/blue, N/S, N/Drugs, athletic, enjoys outdoors, travel. Seeking pretty WPF, 5'2"–5'7", H/W proportionate for romance/LTR.

15. IN KEY

Attractive and active SWM, 34, 5'11", healthy, outdoors type, loves to cook and garden, honest and considerate, open-minded with common sense, enjoys talking.

16. AM I FOR YOU?

SWM, 38, 165 lbs., N/S, N/D, likes outdoors and indoors too. Seeking SF, 20–30.

17. SEEKING SOMEONE SPECIAL

SWM, 32, 5'10", 170 lbs., brown/hazel, enjoys walks, camping, and most outdoor activities. Seeking SWF, 28–35, with similar interests. Kids OK.

18. AVID ADVENTURER

Cute, rugged, strong SWM, 31, 5'11", 175 lbs., college-educated, financially independent, loves snowskiing, boating. Seeking slender woman of grace and sensuality, with inquisitive mind, who shares above interests. N/S, D/D-free.

19. GENUINE NICE GUY

N/S, attractive, DWPM, 34, seeks slender, attractive WPF, 25–40, who is honest and sincere, no games, for summer country concerts, weekend getaways, and romance.

20. WARM AND CARING

Active, athletic, DWPM, enjoys hiking, X-country skiing, dancing, tennis, and quiet times. Seeking similar energy level, female, 40–50, N/S, to share common interests now and into the future.

21. TALL, ATTRACTIVE, CELIBATE

SWM, 41, N/S, N/D, N/Drugs, seeks WF, N/S, 5'9"+, H/W proportionate for friendship and variety of activities; biking, skating, walks, talks, animals, dance, quiet times, possible LTR.

22. IN SEARCH OF

a best friend first. You: WF, H/W proportionate, attractive, pretty, sexy, 25–45. Kids ok. Me: a gentleman, WM, in 50s, ready for LTR.

23. FRIENDS FIRST

Handsome, fun SWM, 50, likes movies, dining out, outdoors. Seeks attractive SWF, 44–50, N/S, light drinker ok.

24. CHRISTIAN FRIEND

Let's be friends first. Native to Seattle. Likes scenic hiking, country fairs, swimming and adventure. I'm 44, never married, seeking attractive SWF, late 20s to early 40s. Let's meet over coffee.

25. BUSTING OUT OF MY SHELL

Sincere, honest good-looking SWM, 31, long dark hair, 5'11", 150 lbs., fit, likes movies, cats, music, creative ideas. Seeking attractive, slender female, 23–26.

26. SEEKING SOMEONE SPECIAL

SWM, 32, 6'1", 185 lbs., seeks SWF, 25–35, who wants to have fun and take a chance on enjoying life.

27. PERRY MASON UNDERSTUDY

Finds legal matters intriguing, kind SWM, seeks compatible female, 45+, with humor. Mutual interests may include: piano music, art, horses, old houses.

28. LOOKING FOR LTR

Old-fashioned, romantic SWM, 23, 5'8", 140 lbs., brown/brown, enjoys bowling, golf, family and the beach. Seeking attractive, outgoing, SW/HF, 21–26, for friendship, possible LTR.

29. FUN-LOVING CHRISTIAN

Handsome, positive DWM, 35, 6'2", physically fit, creative and spontaneous. Seeking an attractive, fun-loving female, 25–36. Kids ok.

30. LONELY IN KENT

SWM, 26, seeks SF, 21–30, interested in outdoor activities, long walks and romantic evenings.

31. SEEKING OLDER WOMAN

Intelligent, good-looking SWM, 31, long hair, 5'10", in great shape, honest, creative. Seeking maturity of attractive, slim woman, 35–55, 130 lbs., who enjoys Christian and secular music, specifically heavy metal. Light drinker/smoker ok. Call me!

32. SUMMER ROMANCE

SWM, 41, 6', 175 lbs., brown/blue, wants summertime love with interesting lady. I have good sense of humor, very open, honest. Are you adventurous?

33. NEW TO WASHINGTON

SWM, 30, funny, open-minded and off-beat, not into looks, age or money. Seeking female companion for fun and romance.

34. LONELY AND SEARCHING

SHCM, 22, 5'5", 120 lbs., seeks SW/HCF, 18–21, with long brown hair, around 5'5" 139 lbs., who enjoys Christian and secular music, specifically heavy metal. Light drinker/smoker ok. Call me!

35. VERY HAIRY GUY
SM, 5'11", 215 lbs., long curly black hair, moustache and hairy body. N/D, N/S, likes mountains, movies, walks, rain. Seeking friendship with female, 35–47.

36. NEED TOUR GUIDE!
Attractive SWM, 32, financially secure. If you like intelligent conversation, dining out, and seek honesty and respect, I won't disappoint you! Seeking attractive, fun, responsible S/DF, 24–35.

37. SINGLE DAD
DWM, 43, 5'9", 185 lbs., loving father of two, handsome, enjoys camping, softball, boating, travel. You: S/DWF, 30-40, N/S, H/W proportionate, caring, honest, spontaneous.

38. TIME'S A WASTIN'
DWPM, 30s, 5'10", H/W proportionate, loves the outdoors. Seeking marriage-minded lady, under 35, N/S, who likes children and appreciates family life and values.

39. RENTON MAN
SWM seeks D/DWF, 29–37, H/W proportionate, who enjoys movies, talking, going out, dinners, etc. N/Drugs, N/D. Smokers ok. Friendship first.

40. LAKESIDE LIVING
DWM, 37, 6'3", fit and trim, N/S, romantic gentleman, jeans or suit, needs compatible south-end type for life's adventures. Kids ok.

41. ROMANTIC COUNTRY BOY
Me: 22, brown/brown, H/W proportionate. You: 22–30, blonde or brunette, H/W proportionate, likes C/W music. Seeking friend and romance for fun. No games.

42. BARBLESS IN SEATTLE
SWM, 39, 188 lbs., H/W proportionate, handsome, N/S, light drinker, loves fly fishing. Seeking SF, attractive, 22–39, H/W proportionate. Must also love fly fishing! Wants LTR!

43. CALL ME NOW
SWCM, 29, 5'10", 190 lbs., brown/brown, enjoys C&W music, movies, camping, sports. Looking for that one special woman to spend time with. N/S.

44. FRENCH MAN SEEKS AMOUR
SWM, 37, N/S, N/D, H/W proportionate, dark brown/green, seeks woman, 27–40, to travel and to explore with. Must be able to enjoy life.

45. SEEKING FUN AND YOU
DWM, 42, 6'3", medium build, enjoys fun, holding hands, walks, travel, camping, boating. Seeking light smoker/drinker. Pinochle a plus. LTR.

46. TOMBOYS AND SINGLE MOMS
Do you like the outdoors, art, tequila, rock music, boating, Mexican food, dancing, romance, adventure, laughter? SWM, 36, N/S, Mercer Island.

47. CLASSY, GQ GENTLEMAN
I'm 46, 5'10", trim, degreed, score an 8 out of 10, enjoy the finer things in life, honest, loving, gentle, and want to love a sweet, beautiful woman, 35–42.

48. ENDANGERED SPECIES . . .
committed father. WM, 43, 5'11", 220 lbs., seeks H/W proportionate mother, N/S, N/D, loves life and outdoors, most of all—complete and committed family life together. South King County.

49. TRUE BLUE CAMPER
SWM, 36, 5'8", 160 lbs., blonde/blue, enjoys hunting, camping, boating, light drinker/smoker. Seeking SWF, 28–40, H/W proportionate, for fun, romance, possible LTR.

50. EXTREMELY NICE GUY
Professional DWM, 47, 5'11", H/W proportionate, kind, considerate, likes drives, bike rides. I am a good catch. Seeking petite, professional female, 35–45, LTR.

51. COURT ME WITH ROMANCE
and experience the unforgettable, SWPF, 30, intoxicating elixir of sensuality and beauty, athletic, adventurous, seeks finely tuned, extraordinary SPM who can accommodate my passion for life.

52. COUNTRY GIRL/CITY STYLE
Cute, petite, DWPF, 37, non-Barbie, seeks tall, non-Ken, must love kids, animals, and fun. N/S, light drinker. Enjoys music, travel, warm nights, no couch potatoes.

53. AGE 22–27?
SCF, 21, pretty, enjoys outdoor activities. Seeking SM with a personality full of character, fun, and depth! Must be Christian, attractive, N/S, N/D.

54. BLONDE, BLUE-EYED
Educated DWF, enjoys beautiful music, dancing, outdoors, seeks DM gentleman, 45–55, who is kind, active, good conversationalist and happy.

55. HIGHLY EDUCATED BEAUTY
Tall, exciting, athletic, natural brunette seeks 4-year degreed (or more), charming business exec, 35–45, 6'+, interested in sports, the outdoors, cultural pursuits and a LTR.

56. ARTISTIC/ACTIVE/
ATTRACTIVE
Petite lady, 59 years young, looks 15+ years younger due to holistic lifestyle, seeks sensitive, secure soul mate, N/S, trim, fit gentleman of intelligence, awareness and sparkle.

57. BIG BEAUTIFUL SWF
Redhead seeks rugged, outdoor type, smoker ok, must like cuddling and dancing. Prefer over 5'11".

58. REDHEAD SWF SEEKS
sincere relationship with tall (6') SWM, enjoys camping, dancing and movies, who wants to have fun. Prefer N/S. N/Drugs, light drinker ok.

59. LOOKING FOR WHAT?
Let's look together. DWF, 50, 5'1", H/W proportionate, N/S, light drinker, family-oriented, likes having fun doing almost anything. Let's have coffee, talk, see what follows, maybe just friends.

60. LIFE WOULD BE COMPLETE
if I had the special someone to share it with. SWF, 35, blonde/blue, H/W proportionate, easy on the eyes, adventurous, passionate, playful, searching for a man that is open-minded and into sharing and caring as I am.

61. QUEEN SIZE
SWF, 27, seeks king size SWM, with good heart. Are you tired of being overlooked because of your weight? Me too, let's talk. Don't be shy!

62. GOLF ANYONE?
Outgoing, classy, athletic. Eastside DWPF, 44, brunette, H/W proportionate, financially secure, loves golf, biking, skiing, theater, fine dining, good conversation. Seeking professional counterpart, 38–48.

63. PRECIOUS JEM
JF, 40, 5'7", 180 lbs., brown/brown, sensual, sensitive, stubborn, enjoys crafts, country living, hobbies, traveling, seek SCM gentle-spirited, 35–43 with similar characteristics for friendship/LTR.

64. PASSIONATE AND STYLISH
Sensual intelligence, sublime touch, and the rhythm of the blues. All with significant sophistication at 31, for an appreciative professional over 45.

65. OLD-FASHIONED LOVE
DWF, 45, N/S, seeks LTR/possible marriage with professional M, with intelligence, integrity, humor, class. Only those commitment-minded please apply.

66. SMART, PERKY WITH ZEST
SWF, H/W proportionate, attractive, east and west coast person, with zest for life, likes boating, outdoor activities, every day is a joy. Seeking professional gentleman, 50s, with same ideas/values.

67. SEEKING A SPECIAL FRIEND
SWF, 47, H/W proportionate, enjoys fishing, camping, outdoors, animals, C/W. Seeking one-woman man, 43–55, light smoker/drinker ok, for friendship first, possible LTR later.

68. A FANTASY?
SWF, 43, 5'6", H/W proportionate, financially secure, active, enjoys life. Seeking SWM, 40–55, H/W proportionate, fit, financially secure, who can dance, N/S, light drinker, disease-free.

69. LEGS, LEGS, LEGS
SWPF, 30, 6'2", slender, looking for you . . . SWM, 28–40, N/S, who likes country music, dinners out, movies, picnics, and finding the humor in life.

70. SEEKING ROMANTIC
COMPANION
DWF, 45, pretty, H/W proportionate, 5'4", N/S, light drinker, enjoys soccer, movies, cooking, cuddling, seeks SWM, 40–50, 5'10"+, attractive, fun-loving, H/W proportionate, sincere.

71. SPIRITUALLY-MINDED
SJPF, 42, pretty, with fun sense of humor, varied interests, H/W proportionate, N/S, light drinker, would like to get to know genuine male 40–48, under 6' for LTR. No games.

72. I'M A BRAT!

SWF, 29, wants to play. Love to camp, golf, bike, travel. Seeking SWM, 27–45, who is financially secure, N/S, can keep with my quick wit. No geeks.

73. LIFE'S BEST WHEN SHARED

Beautiful brunette, New Yorker CEO, 35, 5'4", 105 lbs., caring, intelligent, Kosher, seeks handsome, successful, educated, traditional JM, with strong values; kind, generous, loving.

74. I'M LOST—FIND ME!

DWF, 30, loves sun, dancing, romance, flexible—I'll follow your lead. If you're fit and built like Van Damme, you're my man. No geeks.

75. EASTSIDE CLASSY

Passionate, 39-year-old SWPF, enjoys outdoors, very fit and versatile. Seeking N/S, light drinking, tall, sincere and financially secure WPM for LTR. I'm 5'10", blondish/blue, slender, romantic, pretty, fun!

76. ROCKY MOUNTAIN HIGH

High-energy, attractive SWF, H/W proportionate, professional with passion for conversation, cuddling, sailing, skiing, boating, bicycling, travel and tenderness. Seeking SWM, 55+, H/W proportionate, N/S.

77. R U MAN ENOUGH?

DWF, 38, 5'11", H/W proportionate, enjoys biking, gardening, travel and darts. Seeking tall S/DWM employed, clean-cut, for LTR. Light smoker/drinker, D/D-free.

78. BORN COUNTRY

A heart full of love, a genuine friend, a happy lady! DWCF, N/S, N/D. Very active, secure, slender, young 53.

79. SUMMER FUN

Attractive WF, 39, 5'10", blonde, devoted part-time mom, seeks tall, Eastside gentleman. We're athletic, educated, healthy, N/S, desire LTR, with romance.

80. RSVP ASAP

SWF, 25, easygoing, secure, professional, enjoys boating, scuba diving and traveling. Seeks SWM, 25–35, stable, attractive and financially secure.

81. KIND & COMPASSIONATE

Petite, attractive, SWPF, early 50s, seeks soulmate for LTR. I'm active, spiritual and enjoy outdoor activities, movies, bookstores, travel, animals, and many other interests.

82. TAKE A CHANCE

Very attractive SWF, mid-30s, 5'6", H/W proportionate, blonde/blue, educated, outgoing, introspective, caring, sensual. Seeking same in fun, attractive male, 35–40, 5'10"+, 200 lbs. for possible LTR. N/S, light drinker.

83. CUTE, SWEET & PETITE

Attractive DWF, 43, 5'2", 115 lbs., long light-brown/green, enjoys boating, hiking, walks on beach, trying new things. Seeking someone special 36–47, 5'9"–6'2", physically fit, secure for monogamous LTR.

84. COUNTRY RAISED

SWCF, H/W proportionate, 5'6", 160 lbs., country, lives in city. Seeking SWCM, 40–48.

85. TOTALLY UNIQUE

This SWF, 42, attractive, bright, educated, unpretentious, enjoys country living, nature, walks, animals, gardening, reading, relaxing. Seeking SWM, similar age and interests, for LTR.

86. BLUE JEANS & MINK

DWCF, 5'7", 130 lbs., fit, attractive, family-oriented, spiritually/emotionally strong with varied interests: Mozart to camping, etc. Seeking honest WPM, 40–48 for LTR.

87. ATTRACTIVE BROWN-EYED GIRL

This 40-year-old, H/W proportionate N/S is looking for my best friend, who wants LTR. Must be confident, loyal, honest, emotionally/financially together, with Christian values. Please call. N/S only.

88. CRITIQUE THIS

SWF, 27, outgoing, spontaneous, professional, enjoys boating, jeeping, island hopping. Seeks SWM, 27–37, attractive, confident, financially secure, adventurous, with similar likes.

89. SMILING EYES

Attractive DWF, 47, N/S, light drinker, N/Drugs, outgoing, affectionate, kind, spontaneous, positive, active, professional seeks SWM, 43–50, H/W proportionate 6', with similar values and morals. Loves nature.

90. ROSES ARE RED

Violets are blue. I'm sick of the dating scene, how about you? SWF, 27, sick of game players, losers. Seeking LTR with professional SWM 27–35.

91. FASHIONABLY LATE

SWF, 39, 5'7", athletic, attractive, seeks one special man, intelligent, 6'+, successful, adventurous. Looking for somebody special.

92. WARM, COZY LIFE

DWF, 59, has happy, affectionate, tasteful lifestyle to share with sturdy, appreciative, educated man. Honesty, humor, communication, compassion a must for both. Eastside, Seattle, NS.

93. ENERGETIC

Sophisticated, educated lady, likes outdoors, ethnic activities, sunshine, flowers, life in general. Life's too full for movies or TV. Seeking SWM, 45–55, 5'10"+.

94. LIFE IS GOOD . . .

and getting better! Widowed WF, 53, enjoys family/friends, home/garden, travel/walks. Seeking articulate, metaphysical, romantic SWPM, N/S, light drinker. N/Drugs. Leave your message.

95. ALONE

Attractive DWF, 50+, wants companion/friendship with spontaneous, adventurous S/DWM, 55–60, 5'10"+, N/S, light drinker, with good sense of humor, enjoys outdoors and travel.

96. CUTE, FUN AND JUST ME

Honest, fun-loving, brown/brown, 33, 5'4", H/W proportionate, D/D-

free, N/S, light drinker, seeks same in Harrison Ford/Tom Hanks type to share life's pleasures.

97. PETITE BLONDE

Successful, professional DWF, 44, Christian, N/S, light drinker, "loves water & sun," pretty smile, 5'3", 110 lbs., seeks successful man who loves children and family for fun, romance, LTR.

98. WILL PAINT YOUR WAGON

Scintillating artist, 60s beauty. WPF, stylish fun, gourmet cook, desire savvy SWP "He-man," 55–65, 5'9", N/S, fun loving companionship/life adventures.

99. CARING AND HAPPY

DWF, 54, 5'6", N/S, height/weight proportionate, enjoys golf, movies, dancing, sports. Happy with life, like to meet man with similar interests. Let's meet for coffee.

100. SEEKING PRINCE CHARMING

42 year-old mother (of three boys), 5'2", long blonde/brilliant blue, well-educated, active, altruistic. You? Fun, adventurous, secure, no hang-ups. Integrity, affection, kindness and prayers are musts.

CODE SHEET FOR CODING NEW VARIABLES

Case	SEXUAL IM1	SECUR1	Case	SEXUAL IM1	SECUR1
1			26		
2			27		
3			28		
4			29		
5			30		
6			31		
7			32		
8			33		
9			34		
10			35		
11			36		
12			37		
13			38		
14			39		
15			40		
16			41		
17			42		
18			43		
19			44		
20			45		
21			46		
22			47		
23			48		
24			49		
25			50		

NAME: _____

COURSE: _____

DATE: _____

Workbook exercises and software are copyrighted. Copying is prohibited by law.

EXERCISE

11

WORKSHEET

This worksheet section continues to use the **CONTENT** data file. If you did not immediately continue from the previous section, you will need to open the file and set the "not stated" category to missing data. To do this using the Windows 95 version of the software, select the FILE SETTINGS option from the **FILE & DATA MENU** (if you are using the DOS version of the software, simply click on the button [Set Categories to Missing Data] on the main menu). In one of the blank boxes, type the phrase **Not Stated**, and then click [OK].

1. Ideally, all variables on all cases would be coded at least twice. However, this type of coding takes considerable time. To understand the process, you will code (enter data for) the *first 50* cases on just two variables: 14) SEXUAL IM1 and 17) SECUR1. Look at the special notes on each of these variables in the variable descriptions before beginning your coding. Use the code sheet preceding this worksheet section to code the cases first, then enter the data into the MicroCase data file.

    Note: If you are using the DOS version of Student MicroCase, simply select the ENTER DATA task and enter the data as described in the preliminary section of this exercise. (You will find it easier to navigate between the two variables for which you are doing data entry if you use the mouse to click back and forth between these variables.)

    If you *are* using the Windows 95 version of Student MicroCase, select the ENTER DATA task from the **FILE & DATA MENU.** Click [OK] to accept the default settings as Grid Entry format.

    a. After you have finished the coding, use the Cross-tabulation task to look at the relationship between 5) SEXUAL IMP and 14) SEXUAL IM1.

    How many cases have been coded the same on both
    variables?   _____

                                                        V = _____

    Based on this information, do you think 5) SEXUAL IMP is reliable
    enough to be used in analysis? (Circle one.)   Yes   No

b. Now look at the relationship between 8) SECUR and 17) SECUR1.

How many cases have been coded the same on both
variables?   _____

V = _____

Based on this information, do you think 8) SECUR is reliable
enough to be used in analysis? (Circle one.)   Yes   No

2. When a range is given rather than an age, the younger end is used in the
coding. What other technique(s) could be used for coding the age when a
range is given? Discuss the relative advantages and disadvantages of these
approaches.

3. Test the hypothesis that the sexual implications are more likely to be stressed
in ads placed by males than in those placed by females. Please show your
analysis.

	Data File:	**CONTENT**
	Task:	**Cross-tabulation**
➤	Row Variable:	**5) SEXUAL IMP**
➤	Column Variable:	**1) GENDER**
➤	View:	**Tables**
➤	Display:	**Column %**

	Male	Female
Explicit	_____	_____
Implicit	_____	_____
None	_____	_____

V = _____

Prob. = _____

What is your conclusion?

4. Test the hypothesis, using the original variables, that financial security of the desired respondent will be more important in ads placed by females than in those placed by males.

	Male	Female
Stated	_____	_____
None Ind.	_____	_____

V = _____

Prob. = _____

What is your conclusion?

5. Test the hypothesis that an interest in long-term relationships is more likely to be expressed in ads by females than in ads by males.

	Male	Female
Ltr.	_____	_____
?	_____	_____
No	_____	_____

V = _____

Prob. = _____

What is your conclusion?

6. Test the hypothesis that an interest in long-term relationships is affected by the age of the advertiser.

	Under 40	40 & Over
Ltr.	_____	_____
?	_____	_____
No	_____	_____

V = _____

Prob. = _____

What is your conclusion?

Projects

CREATING A MICROCASE FILE

With Student MicroCase, you can create a file with 50 variables in the Windows 95 version (15 variables in the DOS version) for as many as 100 cases. While you can create only one such file at a time, you can create new files as many times as you want—each new file simply replaces the old file. If you do any of the projects described later, you will need to set up such a file.

Let's use a very small set of hypothetical data to demonstrate how to create a MicroCase file and enter data. We will use just three variables for just five cases, but the same procedures would be used if you were creating a much larger data file.

Suppose we have five students who answered the following mini-questionnaire:

1. Do you live at home, in a dorm, or where?

    1. At home with my parent(s)
    2. In my own apartment or house
    3. In a dorm
    4. In a sorority
    5. In a fraternity

2. I'm not sure that college is worth all the bother.

    1. Strongly agree
    2. Agree
    3. Disagree
    4. Strongly disagree

3. What is your GPA (Grade Point Average)? _____ (write in number)

Each answer to the first two questions has been coded with a number. This is not necessary in the last question, since the response itself will be a number. Generally, when you are entering data for a survey, you will have a questionnaire for each respondent and you would enter data directly from the questionnaire. For example, the responses of the first student are shown below in bold.

1. Do you live at home, in a dorm, or where?

   1. At home with my parent(s)
   2. In my own apartment or house
   3. In a dorm
   4. In a sorority
   5. In a fraternity

2. I'm not sure that college is worth all the bother.

   1. Strongly agree
   2. Agree
   3. Disagree
   4. Strongly disagree

3. What is your GPA (Grade Point Average)? ___**3.6**___ (write in number)

Hypothetical data is provided in the table below for the first five students who were surveyed.

	LIVE?	COL.WORTH	GPA
Student 1	1	3	3.6
Student 2	3	2	3.2
Student 3	2	4	4.0
Student 4		1	2.3
Student 5	1	4	3.0

We go through three steps in order to create a new MicroCase data file and enter the student responses: creating the file, defining the variables that will be included in the file, and entering the actual data for those variables. Let's go through these steps for the student data. Since the steps are different for *Student MicroCase for Windows 95* and *Student MicroCase for DOS*, separate sets of instructions are given.

INSTRUCTIONS FOR WINDOWS 95 VERSION

STEP 1: CREATE A NEW DATA FILE

Important Note: Before you learn how to create a data file, it is important to know a couple of things. First, this version of Student MicroCase permits you to create only one data file at a time. To create a second data file, you must, essentially, replace the first data file you created. Next, if you want to replace a data file you created, DO NOT use the delete key or the recycling bin (or any other Windows operation) to delete this file. This file must be *replaced* within Student MicroCase using the NEW FILE option. For more details on this, see the last part of the Windows 95 instructions below entitled "Replacing a File that You Created With a Newer File."

Go to the **FILE & DATA MENU** in *Student MicroCase for Windows 95*. Select the task named NEW FILE on the menu. On the screen that appears, you will be asked to provide a name and a description for the file that you are creating. For the File Name, type a descriptive name consisting of 1–8 characters. For this example, type **STUDENT** and then press the <Tab> key to move the cursor to the File Description box (if you accidentally pressed <Enter>, you will be asked if you want to enter a variable description). Then type a description of the file, such as the following: **Student Questionnaire**. Then click [OK] to return to the main menu.

STEP 2: DEFINE THE VARIABLES

Each question will become a variable in this data file. So now you will provide information about each variable (or question). Select the DEFINE & EDIT VARIABLES task. The term *undefined* is highlighted. The program is asking for a name for the first variable. *A variable name contains from 1 to 10 characters, including any blank spaces in it*. The name should give you a quick idea of what the variable is. The first question asks where the student lives. Let's call this variable "LIVE?". So where the term *undefined* now appears, type **LIVE?** and press the <Tab> key.

Next, give the variable a description. For this example, type in the first question: **Do you live at home, in a dorm, or where?** and then press the <Tab> key.

The program now asks for the minimum possible value and the maximum possible value for the variable. For the minimum value, type **1** and press the <Tab> key. For the maximum possible value, type **5** and press the <Tab> key. Press the <Tab> key again to skip the question about the number of digits.

Now you can give a 1–10 character label for each value (or category) of the variable. In this situation, code 1 simply stands for "at home with my parent(s)." So type **At home** in the Category Name column, and then press the <Tab> key. For category 2, type **Own place** and press the <Tab> key. For category 3, type **Dorm** and press the <Tab> key. For category 4, type **Sorority** and press the <Tab> key. Finally, for category 5, type **Frat** and press the <Tab> key.

At this point, check the information on the screen for accuracy. If there are any mistakes, you can simply position the mouse cursor wherever you need to in order to correct the mistake.

Next, click the [Continue] button at the top of the screen in order to start work on the second variable. The second variable is a question about the value of college. Let's call this variable "COL.WORTH". So, type **COL.WORTH** and press the <Tab> key. For the variable description, type **I am not sure that college is worth all the bother** and press the <Tab> key.

Next, enter **1** for the minimum value and **4** for the maximum value. Press the <Tab> key again to skip the question about the number of digits. Then type the category name **Str Agree** and press the <Tab> key. Next type the category name **Agree** and press the <Tab> key, then **Disagree** and press the <Tab> key. Finally, type **Str Disagr** and press the <Tab> key.

Click [Continue] to begin work on the third variable, the Grade Point Average. Type **GPA** and press the <Tab> key. For the variable description, type **Grade Point Average** and press the <Tab> key.

Unlike the previous variables, this is not a categorical variable and there is a decimal point in it. Further, in dealing with such variables, it might be difficult to determine in advance what the minimum and maximum values are. So, press the [Tab] key twice to skip the minimum and and maximum values.[1] For the **Decimal Digits**, type **1** to indicate that there is one place to the right of the decimal point.

We have now defined all the variables. So, click [Menu] to exit from this screen and return to the main menu.

STEP 3: ENTERING THE DATA

You are now ready to type in the actual data. Select the ENTER DATA task, select the [Grid Entry] option, and click [OK].

The next screen presents the data entry matrix. Click on the cell at the upper left to position the cursor there. Refer back to the hypothetical data for the five students shown earlier in the table. The first student lives at home (category 1). So, type **1** and press the <Enter> key. He, or she, selected disagree (category 3) as the response to the second question. So, type **3** and press the <Enter> key. The student has a GPA of 3.6. So, type **3.6** and press <Enter>.

We're now ready for the second student. The answer to the first question is 3, so type **3** and press <Enter>. Type **2** for the second variable and press the <Enter> key. Type **3.2** for GPA and press <Enter>.

Although the third student lives in his or her own apartment (category 2), type **6** and press <Enter> to demonstrate a point. The program gives a message that the value is out of range—the value is outside the range defined by the lowest and highest values that we specified. This feature helps to avoid at least certain types of data entry errors. Click [OK] to clear the message, click on the first cell for case 3, and enter **2**. Then enter **4** as the response to the second question and **4.0** as the GPA.

We are now ready for the fourth student. Notice this student did not answer the first question. Simply press <Enter> to leave this variable blank for this case and the program will automatically treat this value as missing. For the COL.WORTH variable, enter **1**, and then enter **3.0** for GPA.

---

[1] While we do not recommend the procedure, you may switch between *Student MicroCase for DOS* and *Student MicroCase for Windows 95* when defining variables and entering data. If, for some reason, you will be entering data in *Student MicroCase for DOS*, you must define a minimum and maximum for the variable.

Do not enter any data for the fifth and final case—we will do that in the next section. For now, click on [Menu] to return to the main menu. The program will now ask whether you want to save the changes to the data. Click [Yes]. That's it. You have created a complete data file and you can now use it the same way that you use any of the other data files in the MicroCase package.

Question Entry

This option will show the question with its possible answers during data entry. The value for the current case is selected by clicking on the appropriate category. Those inexperienced in data entry are less likely to make mistakes using this option.

Using the STUDENT file you created in the previous section, select the ENTER DATA task from the FILE & DATA MENU. Then select the [Question Entry] option, and click [OK]. When the data entry screen appears, the cursor is shown in the grid on the next case and variable. Let's enter the fifth case to our hypothetical data file. First, it may be necessary for you to use the scroll bars located to the right of the grid to position the highlight at the first variable for the fifth case. Go ahead and do that if you need to.

The fifth student lives at home so double-click the **At home** category shown in the entry box at the top of the screen. The value 1 will automatically appear in the appropriate location on the grid below, and the entry screen will advance to the next survey question. This student strongly disagrees with the statement that "I'm not sure that college is worth all the bother" so double-click on the **Str Disagr** category shown at the top of the screen. The GPA variable doesn't have predetermined categories from which to select (because the values will be decimal), so you are prompted to enter a value. Enter **3.0** and press <Enter>. At this point you are advanced to the next case in the data file. We have finished data entry of our five hypothetical cases, so you can return to the main menu by clicking the [Menu] button. You will be asked if you want to save your data—click [Yes].

Rekey Verification

In order to eliminate data entry errors, you might want to reenter the data using the rekey verification process. The underlying assumption of this validation procedure is that the same entry error is unlikely to occur both times a value is entered.

With the STUDENT file open, select the FILE & DATA MENU, select the ENTER DATA task, select the [Enable Rekey Verification] option and one of the format options for data entry, and click [OK]. Now you can reenter the data for each case and each variable. If the value matches the existing value the program continues in the normal fashion. If there is a mismatch, the program will ask you to select the correct entry.

Return to the main menu by clicking the [Menu] button. You will be asked if you want to save your data—click [Yes].

MODIFYING A FILE THAT YOU CREATED

Once you have created a MicroCase data file, you might need to modify it for various reasons (e.g., to correct errors, to add more data, or to add new variables).

Adding More Data or Changing the Data.   If you are correcting data or adding new data (or eliminating some of the data already in the file), then you begin by opening the file and selecting the ENTER DATA task. Select either data entry option and the matrix of data will then appear. You can move about in the data by simply clicking on the spot to where you want to go. You can make any changes you need to make. When finished, click on [Menu] to return to the main menu.

Modifying the MicroCase Data File.   If you are modifying the MicroCase data file (e.g., adding new variables or correcting errors in the variable descriptions), then you begin by selecting the DEFINE & EDIT VARIABLES task. The screen that appears is ready to accept new variables, and you simply continue as before if this is what you are doing. If you are making changes on old variables, then use the arrow buttons that appear below the **Current Variable Number** to scroll to the variable you want to change. For example, if you were making changes to variable 2, then you would use the left/right arrows underneath the box to scroll to the number 2.

When you finish with any additions or modifications, you click the [Menu] button to return to the main menu.

REPLACING A FILE THAT YOU CREATED WITH A NEWER FILE

If you have gone through the preceding example, then you have created a data file named **STUDENT.** Having completed this exercise, however, you are now ready to create a file of your own. How do you replace the STUDENT file with a new file? You simply create a new file (starting with the NEW FILE option) and it replaces the old file automatically. *Do not use the* <Delete> *key to remove the existing file.*

The last point deserves extra emphasis. As noted at the beginning of the "Create a New Data File" section, this version of Student MicroCase permits you to create only one data file at a time. To create a second data file, you must replace the first data file you created using Student MicroCase's NEW FILE option. DO NOT use the delete key or the Windows recycling bin (or any other Windows option for deleting files) to remove this file. If you do, the NEW FILE option will no longer work in Student MicroCase.

If, despite the above warnings, you accidentally delete a file you created, there is a way to reset this file so that the NEW FILE option works again. First, close the Student MicroCase program. Then use the My Computer feature in Windows 95 to locate the data files (e.g., GSS96, USA) for Student MicroCase. If Student MicroCase is accessing your data files from the floppy diskette (such as in a lab setting), then you will find these files on your A or B drive. If you did the standard installation and you installed the file to your hard drive, you will find these files in the C\Program Files\MicroCase\CS directory or folder. Once you have found your data files using My Computer, double-click on the application file named RESET. A brief program will run that resets part of Student MicroCase so that you can again create a data file. Answer **Y** (for Yes) to the prompt warning that a file will be overwritten. You are now returned to the Window desktop (you may have to click the "x" to first close the DOS box that appears when you run the RESET program). Then restart Student MicroCase.

INSTRUCTIONS FOR THE DOS VERSION

STEP 1: CREATE A NEW DATA FILE

From the list of files on the main menu, select the file labeled **NEW_FILE**. Then select the SET UP FILE task.

On the screen that appears, you will be asked to provide a name and a description for the file that you are creating. For the File Name, type a descriptive name consisting of 1–8 characters. For this example, type **STUDENT** and then press the <Enter> key. Then type a description of the file. Here, type the following description: **Student questionnaire**. Then click [OK] to return to the main menu.

STEP 2: DEFINE THE VARIABLES

Now you will provide information about each variable. Select the DEFINE VARIABLES task. The program asks for a name for the first variable. *A variable name contains from 1 to 10 characters, including any blank spaces in it.* The name should give you a quick idea of what the variable is. The first question asks where the student lives. Let's call this variable "LIVE?". So, type **LIVE?** and press the <Enter> key.

Next, give the variable a description. For this example, type in the first question: **Do you live at home, in a dorm, or where?**, and then press the <Enter> key.

At this point, we need to specify whether this variable is categorical or decimal. We will be assigning codes to some predefined categories, so press <Enter> to accept the default categorical setting.

The program now asks for the lowest possible value and the highest possible value for the variable. For the low value, type **1** and press <Enter>. For the highest possible value, type **5** and press <Enter>.

Now you can give a 1–10 character label for each value of the variable. In this situation, code 1 simply stands for "at home with parent(s)." So type **At home** and press the <Enter> key. For category 2, type **Own place** and press <Enter>. For category 3, type **Dorm** and press <Enter>. For category 4, type **Sorority** and press <Enter>. Finally, for category 5, type **Frat** and press <Enter>.

At this point, check the information on the screen for accuracy. If there is any mistake, you can simply position the mouse cursor wherever you need to in order to correct the mistake.

Next, click on the [OK] button at the upper right to start work on the next variable, a question about the value of college. (Or, you can click on the number 2 at the top of the screen.) Let's call this variable **COL.WORTH**. So, type **COL.WORTH** and press <Enter>. For the variable description, type **I am not sure that college is worth all the bother** and press <Enter>.

This is another categorical variable. So, press the <Enter> key again to accept the default categorical variable type.

Next, enter **1** for the low value and **4** for the high value. Then, for category 1, type **Str Agree** and press <Enter>. For category 2, type **Agree** and press <Enter>, and then type **Disagree** and press <Enter>. Finally, type **Str Disagr** and press <Enter>.

Now click on the [OK] button to begin work on the third variable, the Grade Point Average. Type **GPA** and press <Enter>. For the variable description, type **Grade Point Average** and press <Enter>.

Unlike the previous variables, this is not a categorical variable. So, click on **Decimal** to specify that this is a decimal variable and press <Enter>.

For the low value, type **2.3** and press <Enter>. For the high value, type **4.0** and press <Enter>. When we have just a few cases like this, we can easily determine the lowest value and the highest value. However, if we had a large number of cases—or if we had not collected all the data yet—we might not be able to specify easily what the low and high values are. Thus, you would need to make sure to specify a low value that is low enough that no case would go below it and a high value that is high enough that no case would go above it.

We have now defined all the variables. So, click [OK] and then click [Exit] to exit from this screen and return to the main menu.

STEP 3: ENTERING THE DATA

You are now ready to type in the actual data. Select the ENTER DATA task. The next screen presents the first case and it is ready for you to type in the data for the first variable. (You may find it helpful to refer back to the table containing the hypothetical data for the five students.) The first student lives at home (category 1). So, type **1** and press <Enter>. He, or she, selected disagree (category 3) as the response to the second question. So, type **3** and press <Enter>. The student has a GPA of 3.6. So, type **3.6** and press <Enter>.

We're now ready for the second student. The answer to the first question is 3, so type **3** and press <Enter>. Type **2** for the second variable and press <Enter>. Type **3.2** for GPA and press <Enter>.

Although the third student lives in his or her own apartment (category 2), type **6** and press <Enter> to demonstrate a point. The program gives a message that the value is out of range—the value is outside the range defined by the lowest and highest values that we specified. This feature helps to avoid at least certain types of data entry errors. Click [OK] to clear the message, click on the cell, and enter **2**. Then enter **4** as the response to the second questions and **4.0** as the GPA.

We are now ready for the fourth student. Notice this student did not answer the first question. Simply press the <Enter> key to leave this variable blank for this case and the program will automatically treat this value as missing.

Continue to enter the rest of the data as provided in the previously shown table. When you have finished with the last case, click on [Save] to save the data that you have entered. Then click [Exit] to return to the main menu. That's it. You have created a complete data file and you can now use it the same way that you use any of the other data files in the MicroCase package.

MODIFYING A FILE THAT YOU CREATED

Once you have created a MicroCase data file, you might need to modify it for various reasons (e.g., to correct errors, to add more data, or to add new variables).

Adding More Data or Changing the Data.   If you are correcting data or adding new data (or eliminating some of the data already in the file), then you begin by opening the file and then selecting the ENTER DATA task. The matrix of data will then appear. You will need to use the scroll bar to move to the top of the data file. You can move about in the data by simply clicking on the spot to where you want to go. You can make any changes you need to make. When finished, click on [Save] and then click on [Exit] to return to the main menu and click {Yes] to save your changes.

Modifying the MicroCase Data File.   If you are modifying the MicroCase data file (e.g., adding new variables or correcting errors in the variable descriptions), then you begin by selecting the DEFINE VARIABLES task. The screen that appears is ready to accept new variables, and you simply continue as before if this is what you are doing. If you are making changes on old variables, then click the appropriate variable number at the top of the screen. For example, if you were making changes to variable 2, click on **2** at the top of the screen, then make any changes you want to the variable. When you have finished editing or adding variable information, click on the [Exit] button to return to the main menu and click [Yes] to save your changes.

REPLACING A FILE THAT YOU CREATED WITH A NEWER FILE

If you have gone through the preceding example, then you have created a data file called **STUDENT**. Having completed this exercise, however, you are now ready to create a file of your own. How do you replace the STUDENT file with a new file?

First, select the old file that you created (STUDENT). Second, click the SET UP FILE button. Third, delete the old variables and data from the file by clicking the DELETE THIS DATA FILE button. Fourth, proceed to replace the old file name and file description with the file name and file description of the new file that you want to create.

Then click [OK] and you will return to the main menu. At this point, you are ready to define the variables in your new file and then you will be ready to enter the new data.

HUMAN SUBJECTS GUIDELINES

You have probably read about the medical experiment in which men with syphilis were denied penicillin, a known cure, in the quest for additional scientific information. Or the studies of the effect of radiation on individuals who were not informed of the potential risks. While the actual research studies in these incidents undoubtedly are more ambiguous ethically than these headlines suggest, the fact remains that until relatively recently, the ethical treatment of subjects in research was left almost exclusively to the judgment of the researcher. In the 1970s, to protect the public from unethical research, granting agencies started to require that all research must stay within a set of ethical guidelines. To guarantee conformity, each research proposal must be approved by a special committee. Today, virtually all universities and granting organizations require a *human subjects* review for proposed research in which the subjects or participants are humans. (Incidentally, these procedures also protect researchers from lawsuits by unethical subjects.)

Educational projects, such as those suggested in this section, are generally exempt from this process unless the results of the research will be published. Your instructor will inform you if you will need any special clearance before proceeding with an assigned project.

If, at some future time, you wish to conduct your own research, you should review the ethical concerns described in the text for your particular research approach. In addition, you should design your research project using the following suggestions:

- If possible, obtain the informed, voluntary consent of subjects in research. If it is not feasible to obtain consent in advance of the research, then try to obtain consent afterward.

- If there is even a modest degree of risk of harm to the subjects, then it is mandatory that you obtain the informed, voluntary consent of subjects prior to the research.

- Minimize any risk of harm to research participants which might result from the research in any way—from the initial stages of the research to any problems that might occur in the aftermath of the publication of results.

- Take whatever steps are necessary to protect the identity of research participants (e.g., destroying information that might identify participants, disguising the location where the research took place, and presenting results of the study in such a way that no individuals can be identified).

- Do not deceive the subject unless it is the only feasible way to achieve the research objective and adequate provisions have been made to protect the subject from harm.

- Protect the privacy and dignity of the participants.

- When subjects have been used in experimental research, detect and remove any harmful consequences to the subject (e.g., stress).

You may also want to consider the following three aspects which come under particularly close scrutiny during evaluation of the use of human subjects in social science:

1. INVASION OF PRIVACY

This is probably the most common ethical problem in social science research. Before collecting any data, social researchers must inform each participant who will have access to the information and how the information will be used. Steps must then be taken to maintain the promised degree of privacy. This is primarily an issue in survey research, and the text discusses several methods used to maintain the confidentiality of such information. In field research, you must also be careful to protect the privacy of those observed.

When illegal or criminal behavior is studied, the researcher has an additional burden. Social scientists, like newspaper reporters and unlike doctors and lawyers, have no right of confidentiality regarding information and sources. If a list of survey respondents or a tape recording is subpoenaed by a court, a researcher may be held in contempt and punished for refusing to produce the information. In such studies, the researcher should not promise a greater degree of confidentiality than is legally possible. A related problem occurs if the researcher uncovers evidence of continuing criminal activity. For example, a field researcher may find that a particular gang burns buildings on a regular basis and may even have knowledge of future crimes. Failure to report this information could lead to criminal charges against the researcher.

2. STUDIES INVOLVING DECEPTION

In some studies, subjects or participants may not be told the complete truth. This is most common in experimental studies. For example, the experiment described in the text involves a deception: Subjects are led to believe they are listening to other participants when in fact they are listening to a tape recording and, moreover, the taped conversation encourages them to believe that another participant has suffered a seizure. Other types of research may also involve some level of deception. Field researchers may sometimes imply they hold a position when, in fact, they do not. For example, in observing a hospital, a researcher may dress as an orderly, a nurse, or a doctor to be less obtrusive. Even survey researchers may sometimes wish to conceal the exact purpose of their survey.

Deception is a touchy ethical issue. If the same information can be obtained without deception, this is almost always preferable. Most human-subjects committees will weigh the degree and effect of the deception against the potential gain of the study. If deception is used, subjects must be *debriefed* after the study to minimize any consequences of the deception. For example, in the experiment described above, subjects who did not leave the room to seek help may later worry and feel guilty about their behavior. The researchers are responsible for discussing the actual study with subjects so that these feelings do not occur. Unfortunately, no amount of debriefing can erase the subject's knowledge of how he or she behaved in the particular circumstances.

3. Informed Consent and Coercion

Participants and subjects in studies must be informed of the purpose of the study and must be allowed the opportunity to refuse to participate. For example, an instructor cannot *require* any student in a class to complete a questionnaire for a research project—such participation must be voluntary.

When adequate information is provided, most social science research projects are readily approved by the relevant committees.

PROJECT 1: DOING A SURVEY

In this project, you can learn something about students at your college or university. Before you begin, decide on the purpose of your survey. Perhaps you want to see if males and females at your school exhibit gender stereotypes. Or you might want to see if there are sex differences in deviance: Are males more likely to have been picked up by the police than are females? You could determine if students at your school are well informed on certain demographic facts, such as what the world population is. You could see to what extent students' political views match those of their parents. You could even replicate the methodological experiment described in Exercise 10.

After you have selected your topic, design your questionnaire. A variety of possible questions are listed in Appendix C, or you can design your own questions. In the DOS version of Student MicroCase, a data file is limited to 15 variables, so you may ask a maximum of 15 questions. In the Windows 95 version, you can have up to 50 variables.

After you have developed your questionnaire, construct your analysis plan. Which questions will be independent variables and which will be dependent variables? What analysis will you use to answer your research question? Are there possible sources of spuriousness to consider? Any potential intervening variables? It is extremely important to develop your analysis plan before finalizing the questionnaire. This is the only way you can be sure to collect data on all relevant variables. **Warning:** Even very experienced professional researchers have made ridiculous errors of omission in their surveys. For example, a multimillion-dollar longitudinal study of education, following people from ages 16 through 26, failed to ever ask whether respondents had completed college. So, after you are sure you have included everything you will need to know, check again.

You will need to decide whether you will use phone interviews, face-to-face interviews, or mailed questionnaires. If you plan to conduct phone interviews, compose your opening statement. For example, you might say the following:

Hello, this is _____. I'm a social science student at _____.
As a class project, I'm conducting interviews with a sample of students.
Could you spare five minutes to answer some questions?

If you are going to do face-to-face interviews, you will want to work out your approach, perhaps adapting the opening phone statement. If you are going to mail the surveys, you will need to write an appropriate cover letter.

You can now pretest the survey on two or three friends. If necessary, rework the questionnaire. Again review your analysis plan to make sure that you have all relevant information.

Now you can select your sample. Obtain a list of phone numbers or addresses for all students (or perhaps all undergraduates) at your college or university, and randomly select a sample of 50 students. You can then proceed to collect the data.

After the data have been collected, create your MicroCase data set. Enter the data from the questionnaires. Complete your planned analysis and write up the results.

PROJECT 2: DOING A COMPARATIVE STUDY

1. In this project, you'll use information from the *Census of Retail Trade* to test ideas about the effect of the age profiles on retail trade. Before you begin, make sure that you have access to the *Census of Retail Trade*. You may want to examine these reports to get an idea of the type of information available.

Develop two hypotheses about the effect of the age distribution on the consumption of products and services, using states as the units of analysis. For example, you might speculate that states with a high percentage of children will have a higher ratio of spending in grocery stores to spending in restaurants.

After you have developed your hypotheses, you will need to find data on the variables. The age distribution is available from many sources—the *Statistical Abstract of the United States* is probably the most convenient. If necessary, convert this number to a rate, that is, from the number of children under age 12 to the percent of the population under age 12. You can obtain the measures of the dependent variables from the *Census of Retail Trade*. Again convert these to rates.

Next, create a MicroCase data file. (Refer to "Creating a MicroCase File" at the beginning of this section.) Test your hypotheses and write up your results.

2. Create a MicroCase data file of the 25 largest nations. (Refer to "Creating a MicroCase File" at the beginning of this section.) You will first have to determine which nations should be included. You may obtain population figures for each nation from the *Statistical Abstract of the United States*. Develop a hypothesis that can be tested with this data set. For example, you might speculate that the higher the per capita income, the lower the birth rate.

Find appropriate data for each country. Convert the numbers to rates if necessary. Enter the data into your MicroCase data file and test your hypothesis. Write up your results.

3. Select a probability sample of 100 counties or 100 cities in the United States. Collect data to test hypotheses about factors that lead to higher crime rates in counties (or cities). The *County and City Data Book* would be a good source of data.

Convert the data to rates if necessary. Set up a MicroCase data file and test your hypotheses. Write up your results.

PROJECT 3: DOING A FIELD STUDY

Two suggested projects are described here. Since you are observing in public places, you need not obtain permission from those individuals you will observe. However, you will probably want to appear as a "natural" element of the setting. If people know you are taking notes on their behavior, they may not only change the behavior, but also confront you or file a complaint about you. An easy guise for a college student is to pretend to be studying: some reading, some writing, and much thoughtful gazing into space.

When you organize your notes and write your report, develop a central theme. What did you find most interesting? Use your observations to support your interpretation of events.

1. Observe children on playground equipment on several different days. Before you start taking notes, you will want to think about the types of questions you might be able to answer. Which pieces of equipment are most popular? Are these preferences affected by the age or sex of the children? Do children play in groups or by themselves? Are there any age or sex differences in group play? Are there pieces of equipment that encourage cooperation or promote competition among the children? Does the play tend to be segregated by age and sex? Are adults present? To what extent do the adults control the children's behavior? Are some children more popular than others? Are there any "outcasts"? If so, can you determine why this happens?

Remember that adults lurking around playgrounds could appear suspicious. If you plan to observe at a school playground, obtain permission from the principal in advance. You might ask the principal to sign a letter that you have prepared so that you have proper "credentials." If you are observing at a public park, you might want to have a letter from your professor on your college stationery.

2. Observe a fast-food restaurant at different times on different days. As a good field researcher, you should obtain permission from the manager or owner before beginning your observations. If you plan to observe the staff, ask the manager not to reveal your role. As a courtesy, try not to occupy a table when customers need it. If necessary, limit your observations to times when the restaurant is less crowded.

You might choose to observe either the staff or the customers or both. What is the age and sex composition of the staff? Who's in charge? Does informal staff interaction appear to be a function of personal characteristics or of their role in the restaurant? Do there appear to be any conflicts among the staff? If there are, how are these conflicts resolved? Do the staff talk with customers or just fill their orders? What is the attitude of the staff? How do they react when inundated with customers? What do they do when there are no customers?

How does the customer base change over time and over the days of the week? How do individuals who eat by themselves behave? Do they ignore others, perhaps reading newspapers or books, or do they try to engage others in conversation? Are there any sex, age, or occupational differences in this behavior? What sort of people eat together? Aside from family groups, are most groups segregated by age and sex? Which kind of customer eats the fastest? Which kind takes the longest? Do some individuals use this as a social event, or do all customers appear to be there strictly for the food? How do they treat the staff? Are there age, sex, or occupational differences in this treatment?

PROJECT 4: DOING AN EXPERIMENT

1. In this experiment, you can assess the effect of the physical attractiveness of children on adults' responses to them. The independent variable will be photographs of children who differ in attractiveness. A written statement about the child will be attached to the photograph—this description will be the same for all photos. The subjects will then be asked to answer questions about the child.

The first step is to develop the experimental manipulation. Obtain pictures of many children. These should all be the same sex and same apparent age. Now have three or four persons sort the pictures in terms of attractiveness. Record the ranks for each photo. Select two photos—one ranked attractive by all raters and one ranked unattractive by all raters. These two pictures will now be the experimental manipulation.

Develop the description that will accompany the pictures. The following description was adapted from a newspaper account of a child available for adoption:

> Three-year-old *David* is a solemn little *boy* who seems to carry the troubles of the world on *his* shoulders. But *his* mellow, sweet personality makes *him* a favorite of everyone who meets *him*. *He* has an amazing ability to put together Legos and *he's* good at games. *He* loves animals and they seem to love *him*. *He* plays better by *himself* than in groups, though *he* does fine with other children. *He'd* do best as part of a family where *his* quiet spirit won't be lost. (Words in italics have to be changed if pictures of female children are used.)

Now determine how you are going to measure the dependent variable. You might say that you are trying to assess the effectiveness of this type of adoption description and ask the subjects to rate the child on several scales:

Will David be adopted soon?
1____Very likely
2____
3____
4____
5____
6____
7____Very unlikely

How well will David adapt to a new family?
1____Very well
2____
3____
4____
5____
6____
7____Not at all well

How mature is David for his age?
1____Very immature
2____
3____
4____
5____
6____
7____Very mature

Will David have problems in school?
1____Very unlikely
2____
3____
4____
5____
6____
7____Very likely

Be sure to include a question that checks on the effectiveness of the manipulation:

Please rate this child in terms of physical attractiveness:
1____Very unattractive
2____
3____
4____
5____
6____
7____Very attractive

Since you will not be trying to generalize to a population, you may select your subjects in any way that is convenient. You should randomly assign each subject to an experimental condition. You might do this by flipping a coin—assign those with heads to condition 1 and those with tails to condition 2. You should have about 15 subjects in each experimental condition. Simply stop assigning subjects to a condition after you have reached the desired number.

Ask each subject to look at the picture and read the description. You could then interview the subject and fill in the questionnaire yourself. Or you could have the subject fill in the questionnaire.

Then create a MicroCase data set and enter the data. Be sure to create a variable for the experimental condition—simply assign each subject a 1 or 2. You could then determine if there were differences in the dependent variables across the experimental conditions. Be sure to check that the experimental manipulation was effective—subjects perceived one child as more attractive than the other.

ALTERNATIVE EXPERIMENT: Do this same study using homely and attractive dogs.

PROJECT 5: DOING CONTENT ANALYSIS

In the study described in Chapter 6, John C. Merrill reported the results of a content analysis of *Time* magazine stories on Presidents Truman, Eisenhower, and Kennedy. This project is based on that study.

The purpose of this study is to use content analysis to see if different magazines put their own "spin" on the behavior of the president. To obtain the multiple coders needed to check reliability in content coding, you might work with one or two classmates on this project, if your instructor approves. Otherwise, you will need to arrange for at least one coder in addition to yourself.

Select a current event in which the president was involved. Then you can go to the library and select several newspapers to compare. These might be from different parts of the country. Or you could select two news magazines or newspapers from different places on the political spectrum. For example, you might select one "mainstream" magazine such as *Time* or *Newsweek*, and a conservative magazine such as *The National Review* or a liberal magazine such as *The Nation*. Obtain from each publication stories on the selected event. Provide each coder with a copy of the stories, and ask him or her to identify each instance of bias and code it as positive or negative in the following categories:

1. *Attribution bias:* use of a "loaded" verb, such as "barked," "smiled," or "waffled," in place of a neutral verb, such as "said."
2. *Adjective bias:* use of a favorable or unfavorable adjective, such as "disorganized," "boring," "forceful," or "effective." These are subjective, or judgmental, adjectives, in contrast to objective, or neutral, adjectives, such as "blue" sky.
3. *Adverbial bias:* use of a favorable or unfavorable adverb, such as "warmly," "curtly," or "slyly." Frequently, these will be combined with attribution bias, such as in "chatted amiably" or "barked sarcastically."
4. *Outright opinion:* reaching a subjective conclusion rather than stating facts: "His inability to stay focused on the issue has confused his supporters as well as his opponents" or "His forceful presentation put his opponents on the defensive."
5. *Photographic bias:* portraying the president in a positive or negative manner in photographs or cartoons.

In tallying the codes, use only instances of bias on which all coders agree. If one coder cites an instance of negative bias that the other coders ignore, drop that instance. You can then compare the two magazines on each type of bias.

Appendix A:
Student MicroCase
Reference Section

This appendix section provides additional information on using Student MicroCase. If you have not already done so, read the instructions in *Getting Started* at the beginning of this book.

The first part of Appendix A provides information on *Student MicroCase for Windows 95*. Little information is provided here because most of it is available via the on-line help. The second part of Appendix A provides information on *Student MicroCase for DOS*. This section is substantially longer because on-line help is not provided in the DOS version of the program.

STUDENT MICROCASE FOR WINDOWS 95

INSTALLATION

The instructions for installing *Student MicroCase for Windows 95* have been provided in the *Getting Started* section of this workbook. If you have problems installing Student MicroCase, first check to see that your computer has *the minimum system requirements. (Note: Student MicroCase for Windows 95* does not work on computers running Windows 3.0 or 3.1. You should instead use *Student MicroCase for DOS*.)

During the installation, you are asked whether you want the data files (e.g., GSS96, USA) installed to your hard drive. If you choose to not install the data files to your hard drive, you must insert the 3.5" floppy diskette (which contains the data files) every time you run Student MicroCase. If you later change your mind about the data file option you want to use, you will need to reinstall the program.

ON-LINE HELP AND SOFTWARE QUESTIONS

Student MicroCase for Windows 95 offers extensive on-line help. You can obtain task-specific help by pressing the <F1> key at any point in the program. For example, if you are performing a scatterplot analysis, you can press the <F1> key to see the help for the scatterplot task.

If you prefer to browse through a list of the available help topics, select Help from the pull-down menu at the top of the screen and select the Help Topics option. At this point you will be provided with a list of topic areas. Topic areas are represented by graphics of a closed book. To see what information is available in a given topic area, double-click on a book to "open" it. (For this version of the software, use only the "Student MicroCase" section of help; do not use the "Student ExplorIt" section.) To open a book, double-clicking on the Standard Operations book, provides a list of the help topics available for standard operations—such as selecting variables and opening files. A help topic is represented by a piece of paper with a question mark on it. Double-click on a help topic to view it.

If you have questions about *Student MicroCase for Windows 95*, the first step is to try the on-line help described above. If you are not very familiar with software or computers, a second option is to ask a classmate or your instructor for assistance. If you are still unable to get an answer to your question or to resolve a problem, go to the Technical Support section of MicroCase's web site at http://www.microcase.com/student/support.html.

STUDENT MICROCASE FOR DOS

If you have not already done so, read the instructions in *Getting Started* at the beginning of this book. The remainder of this appendix section provides additional information on using *Student MicroCase for DOS*. (If you are using *Student MicroCase for Windows 95*, refer to the section above.)

GETTING YOUR MOUSE WORKING

If you are using *Student MicroCase for DOS* on a Windows 3.0/3.1 computer and the mouse arrow fails to appear on your screen, follow these instructions for loading the mouse driver:

1. Exit Windows. (You can exit by using <Alt> <F4> to close each window.)

2. From the DOS prompt (C:\>) type: **MOUSE** <ENTER>

3. If you get an error message (i.e., "Bad command or file name," "Invalid directory") try either: **C:\MOUSE\MOUSE** <ENTER> or **C:\LMOUSE\MOUSE** <ENTER>

4. If you get back a message that tells you the mouse driver has been loaded, type **A:MC** and press <ENTER> to start the program. If the mouse works now, you will need to follow the same procedure for loading the mouse driver each time you want to run *Student MicroCase for DOS*.

5. If you still cannot see the mouse on the screen, you will need to contact the manufacturer of your computer to obtain a DOS mouse driver.

6. To return to Windows, type: **WIN** <ENTER>

HELP AND SOFTWARE QUESTIONS

If you encounter problems with *Student MicroCase for DOS*, the first step is to review relevant sections in this workbook that might address your problem, such as this appendix section. Next you should try the help instructions provided at the bottom of most screens. If you are not very familiar with software or computers, a third option is to ask a classmate or your instructor for assistance. If you are still unable to get an answer to your question or to resolve a problem, go to the Technical Support section of MicroCase's web site at http://www.microcase.com/student/support.html.

STUDENT MICROCASE: DOS VS. WINDOWS 95

There are a few features included in *Student MicroCase for Windows 95* that are not included in *Student MicroCase for DOS*. However, if you use *Student MicroCase for DOS*, these differences will have little impact on you since most relate to data management types of tasks. Just so you are aware, the major menu options that are included in the Windows 95 version of Student MicroCase but are not in the DOS version include: COLLAPSE VARIABLES, RECODE VARIABLES, CODEBOOK, FILE SETTINGS, and FILE NOTES. Also, you should be aware that *Student MicroCase for DOS* operates from a single main menu while the Windows 95 version has two main menus. So ignore any instructions in the workbook that suggest that you should switch to a different main menu.

STANDARD OPERATIONS

These operations appear on the top row of buttons in *Student MicroCase for DOS* (not all buttons appear on all screens):

EXIT

When you are within a task, clicking the [Exit] button will return you to the variable selection screen for the task. If you are on the variable selection screen within a particular task, this button will return you to the main menu. If you are on the main menu, you can exit the program by clicking the [Exit] button.

PRINT

To print the information shown on the screen, click the [Print] button. A window will open giving the current settings for the printer. If the settings are correct, click

[OK] to send the results to the printer. If you change your mind about printing something, click [Cancel]. If the printer settings are incorrect, click [Settings] and follow the instructions in the next paragraph. If you want to save results to a disk file and print it later, use the [Disk File] option. In this case, type in the directory and file name you want to use (using standard DOS conventions) and click [OK]. This option will save text screens, but not graphics screens, to the named file. Also use the [Disk File] option if your printer is not supported by *Student MicroCase for DOS*. Such files can be opened and printed using any word processor program.

Changing Printer Settings. To change the printer settings, click [Print] and then choose [Settings]. To select a new printer from the list, click on its name and then click [OK]. If your printer doesn't appear on the list, try each printer and use the one that works best. The [Text (ASCII)] option will not allow you to print graphics.

You are also given the option of directing the print output to a particular port (the physical connector from your computer to your printer). Use LPT1: unless you have been instructed otherwise. (If you are printing over a network and are unable to print, check with your instructor or network administrator for additional information.)

VARIABLES

The [Variables] button will let you look at the list of variables for the data file that is currently open. The description of the highlighted variable is shown in a box at the lower right of the window. You can move this highlight through the list of variables by clicking on the scroll buttons to the right of the list box, by using the up and down cursor keys (as well as the <PgUp>, <PgDn>, <Home>, and <End> keys), or by clicking a variable once with the mouse.

You can also search the variable names and descriptions for a word, a partial word, or a phrase. Perhaps, for example, you want to find a variable about income. Click the [Search] button. Type **income** and press <Enter> (or click [OK]). The variable list window now contains only those variables which have the word income in either the variable name or the variable description. To return to the full list of variables, click [Full List].

SELECTING A SUBSET

Most statistical analysis tasks in Student MicroCase allow you to analyze a subset of cases. For example, if you wanted to look only at females in the GSS, you would need to select a subset.

To limit your analysis to a subset of cases, select the primary variables for analysis as usual. But before you click [OK] to see the final results, click the [Subset Variables] button. A subset selection screen will appear and you may select up to four subset variables. You should select a subset variable as you would a regular variable. But upon choosing a variable, you will immediately be prompted to provide information on the categories to be used in selecting cases for the subset. Since there are two different types of variables, there are two different ways to select subset categories.

SELECTING SUBSETS WITH CATEGORICAL VARIABLES

A variable is a "categorical variable" when each of its categories has a discrete name or label (e.g., "Male" or "Female" for the variable SEX). After you select a subset variable, the screen will ask you to select a category for a subset. Click it once to move the highlight to it and then click a second time to select the category an "x" will appear in the box to the left of the name. You may select as many or as few categories as you wish.

After you have selected the categories, you must indicate whether cases in the indicated categories are to be included or excluded in the analysis. For example, if you select SEX as the subset variable and "Male" as the category, you may select to "include" only males in the analysis, or you may choose to "exclude" them and look only at females ("non-male"). Just click the option you want. You then click [OK] to return to the variable selection screen.

SELECTING SUBSETS WITH NONCATEGORICAL VARIABLES

If a variable uses a range of numbers for its values (1.2-13.3), and these values do not fall into discrete categories (like "male" or "female"), then a different method of selecting cases is used. After you select the subset variable, the screen will ask for the low and the high value to be used in selecting the cases. For example, if you type 1.2 as the low value and 6.7 as the high value, cases with values between 1.2 and 6.7 on the subset variable will be included in the analysis. You then click [OK] to return to the variable selection screen.

MULTIPLE SUBSET VARIABLES

If you want to base the subset on more than one variable, then you would select a second variable and define another subset. For example, if you wanted to include only white females, you would define your subset using two subset variables: RACE (include category white) and SEX (include category female).

DELETING SUBSETS

All variable selections including selections of subset variables are saved by ExplorIt once you exit from a results screen. If you are going to conduct another analysis using the same task, it is important to delete or clear subset variables when you are done with them. There are three ways to accomplish this. First, you can click the [Clear All] button on the variable selection screen, which will clear all selected variables. Second, you can click the [Subset Variables] button and then click [Delete] to eliminate each the subset variable individually. Or, third, you can return to the main menu, which will automatically deletes all variable selections.

UNIVARIATE STATISTICS

SELECTING VARIABLES

Select a primary variable according to the variable selection instructions in *Getting Started* at the beginning of this book. If you want a subset of cases, see the Selecting Subsets section toward the beginning of this appendix. (If you need to erase all existing variable information for this task, click the [Clear All] button.)

Once your variables have been selected, click [OK] to obtain the results of your analysis.

VIEWS

These options are located in the top row of buttons on the results screen.

Pie. This shows the distribution of the variable and displays it in the form of a pie chart.

Bar [Frequency]. This displays the distribution of cases in the form of a bar graph. Information on the first category of the variable is shown below the bar graph. To see information on other categories, point and click on the bar of your choice.

Bar [Cumulative]. This displays the cumulative percentage of cases for the category displayed.

Statistics. This button produces the frequency, percent, cumulative percent, and z-score for each category of the variable. Summary statistics are shown at the top of the screen.

CROSS-TABULATION

SELECTING VARIABLES

Select a row variable and a column variable according to the variable selection instructions in *Getting Started* at the beginning of this book. If you want a subset of cases, see the instructions on selecting subsets at the beginning of this appendix. (If you wish to erase all existing variable information for this task, click the [Clear All] option.)

Control Variable (optional). The [Control Variables] option can be selected from the variable selection screen in the CROSS-TABULATION task. A window for selecting control variables will open. At this point you may select up to three control variables using the usual method of selecting variables. When you have selected the control variables for your analysis, click [OK] to return to the variable selection. When the final results are displayed, a "partial table" is shown for each level of the control variable(s).

Once your variables have been selected, click [OK] to complete the analysis.

VIEWS

Tables. The resulting table shows the cross-tabulation results for the selected row and column variables. If one or more control variables have been selected, the categories of the cases contained in this partial table are shown at the top of the screen. If the entire table will not fit on the screen, use the cursor keys or click the scroll bars at the bottom and/or right of the table to scroll additional rows and columns.

DISPLAY (these options are located on the menu at the right)

Column %: This option provides the column percentages for the table. Missing data is ignored.

Row %: This option provides the row percentages for the table. Missing data is ignored.

Total %: This option provides the total percentages for the table. Missing data is ignored.

Freq.: This option returns the table to the original frequencies.

Expected: This option provides expected cell frequencies based on the chi-square

Previous (Optional): If a control variable has been selected, this option will let you return to previously viewed subtables.

Next (Optional): If a control variable has been selected, this option will take you to the next subtable.

Stats. This view shows the summary statistics for the table.

Bar. This shows a stacked bar chart representing the table.

Collapse. The [Collapse] button appears whenever a cross-tabulation table is showing. This option allows you to combine categories of a table or to drop categories entirely. (All changes using the [Collapse] option do not permanently modify the variable—the changes disappear as soon as you leave the results screen.) To select the categories to be combined or dropped, click on the category labels (this causes the entire row or column to be highlighted). Then click the [Collapse] button. Here you are given the choice of creating a new collapsed category or turning the highlighted categories into missing data. To create a new collapsed category, type in a category name and click [OK]. If you want to drop the highlighted categories, click the [Drop] button. The table will then be updated according to your selections. The [Collapse] option will work with both the row and column variables in a cross-tabulation table, although only one variable can be modified at a time.

MAPPING

SELECTING VARIABLES

Select a variable according to the variable selection instructions in *Getting Started* at the beginning of this book. If you want to use a subset of cases, see instructions on selecting subsets at the beginning of this appendix. If you wish to erase all existing variable information for this task, click the [Clear All] option.

Once you select your variables, click [OK] to obtain the map.

VIEWS

These options are located in the top row of buttons.

Map. The selected variable is mapped into five different levels from light (lowest) to dark (highest). Cases for which data are unavailable are left blank.

Display

> *Legend*: This option provides the values represented by each level, known as the map legend, in a box at the lower right. Selecting this option a second time removes the legend.

> *Find*: This option allows you to locate a particular case. When you select this option, a list of all cases will be shown. Click twice on the case you want to select so that an "x" appears to the left of the name. If the case you want to select is not shown in the window, scroll to it using the cursor keys or click the scroll bar at the right side of the window. Click [OK] to continue. The selected case is shown in a highlighted color on the map and its name, value, and rank are shown at the bottom of the screen. To deselect this highlighted case, click the [Find] button again or click outside the border of the map.

Special Feature: Point and click on a case: You may use the mouse to point and click on a case. The selected case is shown in the highlight color and its name, value, and rank are shown at the bottom of the screen. To deselect this case, click outside the border of the map.

Spot: This view shows each case as a filled circle. The size of the spot, or circle, is relative to the value of the case—cases with higher values have larger spots. The color of the spot is the same as in the Map view. To return to the original map, click the [Spot] button again.

List: Rank. This option allows you to rank the cases on the variable from high to low. The name of the variable, its rank, and its value are displayed. In addition, the map color for the case is shown to the left of the name. Since the initial screen shows only the highest cases, use the scroll bar at the right side of the window to see additional values.

List: Alpha. This option works the same as [List: Rank], except that cases are shown in alphabetical order.

SCATTERPLOT

VARIABLE SELECTION

Select two variables according to the variable selection instructions provided in *Getting Started* at the beginning of this book. The dependent variable will be represented by the Y-axis on the graph and the independent variable will be represented by the X-axis.

If you want a subset of cases, see instructions on selecting subsets at the beginning of this appendix. (If you need to erase all existing variable information for this task, click the [Clear All] option.)

Once your variables have been selected, click [OK] to obtain the scatterplot results.

VIEWS

The scatterplot task has only one view.

The scatterplot graph shows each case as a dot with the X-axis representing the independent variable and the Y-axis representing the dependent variable. The correlation coefficient (r) is shown at the lower left of the screen. One asterisk indicates a .05 level of statistical significance, two asterisks, a .01 level.

Display. (these options are located on the Display menu at the left of the screen):

Reg. Line: This option places the regression line on the scatterplot graph and gives the equation for the regression line below the graph. Selecting this option a second time removes the line.

Residuals: The residual for each case—a vertical line from the case to the regression line— is shown on the scatterplot graph. Selecting this option a second time removes the residual lines.

Find Case: This option allows you to identify—place a box around—the dot representing a particular case. When you select this option, a list of all cases will be shown. If the case you want to see is not immediately visible in the window, scroll to the appropriate case using the cursor keys or click the scroll bar at the right side of the window. To select a case, click on it twice so that an "x" appears in the box to the left. Click [OK] to return to the scatterplot graph. Information on the selected case is shown in the window at the lower left—this window is described under the special features section below.

Outlier: This option will identify—place a box around—the dot representing the outlier. In this application, the outlier is defined as the case which would make the greatest change in the correlation coefficient if it were removed. When you select this option, information on the outlier case is given in a window at the lower left—this window is described under special features below.

> <u>*Special Feature*</u>: Point and click on dot: You may use the mouse to point and click on a dot. A box will be placed around this dot and information on the case is shown in the window located at the lower left.
>
> X button: If you point and click on the X button at the bottom of the graph, the description of the independent variable will be shown below the graph. Click the "x" in the upper right corner of the description to close this window.
>
> Y button: If you point and click on the Y button at the left side of the graph, the description of the dependent variable will be shown below the graph. Click the "x" in the upper right corner of the description to close this window.
>
> Window: When certain displays are selected from the scatterplot screen, a window will appear that contains information on the case currently highlighted in the scatterplot. This window will appear when you use the [Outlier] option, the [Find case] option, or if you

point and click on a dot in the scatterplot. The name of the case and its value on the independent (or X) variable and on the dependent (or Y) variable are shown. In addition, you are given the option of removing the case from the scatterplot. If you click on the option [Remove Case from Graph], the case will be removed from the scatterplot and any statistics will be recalculated with that case removed. (The removal of this case is only temporary—the original data set has not been modified.) The bottom line of the window tells you what the value and statistical significance of the correlation coefficient will be if the highlighted case is removed. Click on the "x" in the upper right corner of this window to close it.

CORRELATION

VARIABLE SELECTION

Select two or more variables according to the variable selection instructions in *Getting Started* at the beginning of this book. If you want a subset of cases, see instructions on selecting subsets at the beginning of this appendix. (If you need to erase all existing variable information for this task, click the [Clear All] option.)

Once the variables have been selected, click [OK] to obtain the results of your analysis.

VIEWS

The correlation task has only one view. If a correlation value (r) is statistically significant, it is followed by one asterisk (.05 level) or two asterisks (.01 level).

Note: the Windows 95 version of Student MicroCase allows you to toggle between pairwise and listwise deletion; you can also control whether a one-tailed or two-tailed test of significance is used.

ANALYSIS OF VARIANCE (ANOVA) AND *t*-TEST

VARIABLE SELECTION

Select a dependent variable (usually an ordinal/ratio variable) and an independent variable (usually a categorical variable) according to the variable selection instructions provided in the *Getting Started* at the beginning of this book. If you want a subset of cases, see instructions on "Selecting Subsets" at the beginning of this appendix. (If you wish to erase all existing variable information for this task, click the [Clear All] button.) Once the variables have been selected, click [OK] to obtain the results of your analysis.

The resulting graph shows the partitioning of the total variation of the dependent variable. The mean for each category of the independent variable is represented by a small horizontal bar; the standard deviation for each category is shown using a rectangular box. The overall mean is also provided on the right side of the screen.

The *t* TEST task works the same as the ANOVA task except that it is intended to be used in analyses where the independent variable has only two categories. In such cases, the t value is provided instead of the F value.

VIEWS

ANOVA. This view provides statistical summary information for the analysis, including Eta Square, the sum of squares, degrees of freedom, mean square, F (or t), the probability value, and more.

Means. This view provides the means and standard deviations by categories of the independent variables. The probability that the difference of the means is statistically significant is also shown.

REGRESSION

VARIABLE SELECTION

Select a dependent variable and one or more independent variables according to the variable selection instructions provided in the *Getting Started* at the beginning of this book. If you want to use a subset of cases, see the "Selecting Subsets" section toward the beginning of this appendix. (If you wish to erase all existing variable information for this task, click the [Clear All] button. To delete only a previously selected independent variable, click on the variable name and press the delete or backspace key on your keyboard.) Once the variables have been selected, click [OK] to obtain the results of your analysis.

VIEWS

Graph. The graph shows the effect of one or more independent variables on a ratio or ordinal dependent variable. The individual correlations (r) between the independent variables and the dependent variable are shown, as are the beta values. If a beta value is statistically significant, it is followed by one asterisk (.05 level) or two asterisks (.01 level). Multiple R-squared is shown at the top right corner of the screen.

ANOVA. This option provides analysis of variance information including eta square, standardized and unstandardized betas, the sum of squares, degrees of freedom, mean square, F, the probability value, and more.

Correlation. This view provides a correlation matrix of all variables included in the analysis. If a correlation is statistically significant, it is followed by one asterisk (.05 level) or two asterisks (.01 level).

Means. This option provides the means and standard deviations for the variables included in the analysis.

SET UP FILE, DEFINE VARIABLES, AND DATA ENTRY

The SET UP FILE task on the main menu allows you to create an entirely new data file. Once the data file is created you use the DEFINE VARIABLES and ENTER DATA tasks. Because of the many steps involved in creating a new data file and entering data, "Creating a MicroCase File" (in the *Projects* section) discusses these tasks in greater detail.

LIST DATA

The LIST DATA task on the main menu lets you list the data for the TEST and CONTENT data files, as well as a data file that you may create. All cases and variables in the data file are listed in the form of a spreadsheet. Use the scroll bars to view variables or cases not shown on the screen.

Appendix B: Variable Names and Sources

Note: the full version of the MicroCase program can access the data files provided with this book. However, if you use the full version of MicroCase to move variables from these files into other MicroCase files, or vice versa, you may need to re-order the cases. Also, note that files which have been modified in the full version of MicroCase will not function properly in Student MicroCase.

◆ **SHORT LABEL: ANES96** ◆

1) LIKE CONG?
2) WHO IN 92?
3) LIKE 1
4) LIKE 2
5) LIKE 3
6) LIKE 1*
7) LIKE 2*
8) LIKE 3*
9) LIKE 20
10) LIKE 24
11) LIKE 25
12) CLINTON 1
13) LIB/CON
14) FUT_LIVING
15) ISOLATE
16) PARTY
17) SPENDING?
18) MED. COST
19) JOBS
20) HELP BLACK
21) ABORTION

22) CRIME
23) ENV./JOBS
24) EQUAL
25) WHO VOTE
26) EFFECT GOV
27) REL. IMP?
28) REL. GUIDE
29) PRAY
30) READ BIBLE
31) FUNDAMENT.
32) CHRIST TYP
33) BORN AGAIN
34) VOTED?
35) WHO VOTE?
36) FOLLOW GOV
37) IMP GOV
38) GOV SIZE
39) GAY LAW
40) SC. PRAYER
41) FEAR CRIME
42) GUN LAW?

43) POLS CARE
44) FAM VALUES
45) CROOKED?
46) TRUST GOV
47) TRUST
48) TRUST NEWS
49) POLITKNOW
50) POLPART
51) RELIGION
52) USREPVOTE
53) WELFARE
54) AGE CATEGR
55) EDUC.LEVEL
56) FAMINCOME
57) TVSPORTS
58) SEX
59) RACE
60) REGION
61) BELT CODE
62) INCOME CAT
63) COLLEGE?

◆ SHORT LABEL: GSS93 ◆

1) URBAN?
2) PLACE SIZE
3) HUNT/FISH
4) INCOME @16
5) DRINK?
6) WATCH PBS?
7) HIT CHILD
8) HIT BEATER
9) HIT ROBBER
10) HIT OK?
11) BIG BAND
12) BLUEGRASS
13) CNTRY/WEST
14) MUSICALS
15) CLASSICAL
16) OPERA
17) BLUES
18) GOSPEL

19) JAZZ
20) RAP MUSIC
21) HVY METAL
22) ATTNDSPORT
23) VISIT ART
24) AUTO RACE
25) GARDEN
26) DO SPORTS
27) VEGETARIAN
28) MAN MADE
29) ALL DIE
30) CANCER
31) ORGANIC
32) MOM WORK?
33) PLACE SIZ!
34) AGE AT WED
35) VETERAN?
36) SAT.HEALT!

37) PUB.DECIDE
38) BUS.DECIDE
39) SEX
40) RACE
41) WH/AFRI.AM
42) REGION
43) AGE
44) ED YEARS
45) INCOME
46) R.INCOME
47) FEAR WALK
48) DEGREE
49) $ 50%50%
50) OWN HOME?
51) READ PAPER
52) CH.ATTEND

◆ SHORT LABEL: GSS96 ◆

1) SEX
2) RACE
3) WORKING?
4) VOTE IN 92
5) WHO IN 92?
6) ENVIRON. $
7) HEALTH $
8) BIG CITY $
9) CRIME $
10) BLACK $
11) WELFARE $
12) ENVIRON.$2
13) HEALTH $2
14) BIG CITY$2
15) CRIME $2
16) BLACK $2

17) WELFARE $2
18) RELIGION
19) INTERMAR?
20) RACE SEG.
21) EDUCATION?
22) FED.GOV'T?
23) SUP.COURT?
24) CONGRESS?
25) MILITARY?
26) RELIGION?
27) EVER UNEMP
28) WOMEN HOME
29) WOMAN PREZ
30) MEN BETTER
31) FEAR WALK
32) COMPREHEND

33) ATTITUDE?
34) WH/AFRI.AM
35) MARITAL
36) PHONE
37) HAPPY?
38) # SIBS
39) SPANK?
40) ABORT DEF
41) ABORT WANT
42) ABORT HLTH
43) ABORT NO$
44) ABORT RAPE
45) ABORT SIGL
46) ABORT INDX
47) ATHEIST SP
48) RACIST SPK

49) COMMUN SPK
50) MILITI. SP
51) GAY SPEAK
52) FREE SPEAK
53) READ PAPER
54) OWN HOME?
55) SOC. BAR
56) EVER STRAY
57) SEX FREQ
58) # CHILDREN
59) GOV.MED.
60) MUCH GOV'T
61) HEALTH
62) ZODIAC
63) MOVERS
64) DEGREE
65) WATCH TV
66) AGE
67) OVER 50

68) HELP HUSB
69) HOUSEWIFE
70) PRAY
71) HOW RELIG?
72) CH.ATTEND
73) ED YEARS
74) POL.PARTY
75) POL. VIEW
76) INCOME
77) R.INCOME
78) $ 50%50%
79) DAD PREST
80) PLACE SIZE
81) URBAN?
82) REGION
83) R.FUND/LIB
84) AFTERLIFE?
85) SCH.PRAY.1
86) BIBLE1

87) MEDIA REL
88) RELIG. ID
89) DENOM.
90) # SIBS!
91) AGE!
92) EDUCATION!
93) DAD EDUC!
94) MOM EDUC!
95) INCOME!
96) R.INCOME!
97) POL.PARTY!
98) POL. VIEW!
99) CH.ATTEND!
100) HOW RELIG!
101) SEX FREQ!
102) HEALTH!
103) DEGREE!
104) PRAY!
105) INCOME2

◆ SHORT LABEL: USA ◆

1) STATE NAME
2) POP 1990
3) POP GO 90
4) % WHITE
5) % BLACK
6) % ASIAN
7) %N.AMERICA
8) % HISPANIC
9) MEXICAN K
10) P.RICAN K
11) CUBAN K
12) IMMIGRANTS
13) AGE 5–17
14) % OVER 65
15) AVER. AGE
16) % RURAL
17) % METROPOL

18) DENSITY
19) CROWDED
20) MARRIAGE
21) DIVORCE
22) %DIVORCED
23) %M.DIVORCE
24) %F.DIVORCE
25) COUPLES
26) % FEM.HEAD
27) MALE HOMES
28) %SINGLES
29) %SINGLE M
30) %SINGLE F.
31) % WIDOWS
32) % WIDOWERS
33) TEEN MOMS
34) % FEM.WORK

35) ABORTION
36) ADOPTIONS
37) WARM WINTR
38) ELEVATION
39) AREA
40) SOUTHNESS
41) SO.ACCENTS
42) WESTNESS
43) REGION
44) COKE USERS
45) DRUG ED
46) ALCOHOL
47) % WINE
48) % BEER
49) HEALTH IND
50) AIDS DEATH
51) % FAT

52) SUICIDE
53) % FEM MD
54) MDs
55) PLASTIC
56) SHRINKS
57) CHIROPRACT
58) PLAYBOY
59) #PLAYBOY
60) MOTH.JONES
61) N.R./NAT.
62) PEACE CORP
63) %FEMALE LG
64) ART $ PER
65) ARC.DIGEST
66) CONNOISS'R
67) GOURMET
68) COSMO
69) ROLLING ST
70) PICKUPS
71) FLD&STREAM
72) HUNTING
73) FISHING
74) VETERANS
75) PUBLIC AID
76) HOMELESS
77) WELFARE $
78) NEW HOMES
79) % ON AFDC
80) FOODSTAMPS
81) $ PER CAP.

82) HOME VALUE
83) RENT
84) P.TAX/CP
85) AUTOS PER
86) %POOR
87) % UNEMPLOY
88) $ WORKERS
89) % HIGH SCH
90) % COLLEGE
91) DROPOUTS
92) MATH SCORE
93) $PER PUPIL
94) STU/TEACH
95) GRAD ED
96) $ HIGH ED
97) % HIGH ED
98) COLLEGE $
99) LIBRARIES
100) BOOK $
101) %NO RELIG.
102) % JEWISH
103) % CATHOLIC
104) % BAPTIST
105) CHURCH MEM
106) CRIME RATE
107) VIO.CRIME
108) PROP.CRIME
109) MURDER
110) RAPE
111) ROBBERY

112) ASSAULT
113) BURGLARY
114) LARCENY
115) AUTO THEFT
116) %BUSH 1988
117) %CLINTON92
118) %BUSH '92
119) %PEROT 92
120) STATES '92
121) FEM.LEGIS
122) LOBBYISTS
123) BUSH88
124) CLINTON92
125) FS$/PER
126) FS$/CAP
127) MILES/DRV
128) MILES/VHCL
129) CARS/HSE90
130) %CLINTON96
131) %DOLE 96
132) %PEROT96
133) STATES '96
134) POP 96
135) HLTH INS
136) SCHOOLS
137) NDM 95–96
138) CRIMES
139) HISPANICS
140) CRIMERATE

◆ SOURCES ◆

ANES96

The ANES96 data file is based on selected variables from the 1996 American National Election Study provided by the National Election Studies, Institute for Social Research at the University of Michigan and the Inter-university Consortium for Political and Social Research. The principal investigators are Steven J. Rosenstone, Warren E. Miller, Donald R. Kinder, and the National Election Studies.

GSS93

The GSS93 data file is based on selected variables from the National Opinion Research Center (University of Chicago) General Social Survey for 1993, distributed by The Roper Center and the Inter-university Consortium for Political and Social Research. The principal investigators are James A. Davis and Tom W. Smith.

GSS96

The GSS96 data file is based on selected variables from the National Opinion Research Center (University of Chicago) General Social Survey for 1996, distributed by The Roper Center and the Inter-university Consortium for Political and Social Research. The principal investigators are James A. Davis and Tom W. Smith.

USA

The data in the USA file are from a variety of sources. The variable description for each variable uses the following abbreviations to indicate the source.

ABC: Blue Book, Audit Bureau of Circulation

CENSUS: The summary volumes of the 1990 U.S. Census

CHRON.: *The Chronicle of Higher Education Almanac*

CHURCH: Churches and Church Membership in the United States, Glenmary Research Center

DIGEST: *Digest of Education Statistics*, U.S. Dept. of Education

E&E: *Employment and Earnings*, U.S. Bureau of Labor Statistics

HCSR: *Health Care State Rankings*, Morgan Quitno

HIGHWAY: *Highway Statistics*, U.S. Department of Transportation

KOSMIN: Kosmin, Barry A. 1991. *Research Report: The National Survey of Religious Identification*, New York: CUNY Graduate Center

MMWR: *Morbidity and Mortality Weekly Report*, Centers for Disease Control

MVSR: *Monthly Vital Statistics Report*, Centers for Disease Control

S.A.: *Statistical Abstract of the United States*

SMAD: *State and Metropolitan Area Data Book*, 1991, U.S. Dept. of Commerce

S.P.R.: *State Policy Reference*

SR: *State Rankings*, Morgan Quitno

UCR: *The Uniform Crime Reports*, U.S. Dept. of Justice

WA: World Almanac

Appendix C: Student Questionnaire

Please answer the following questions by placing a mark in the appropriate blank or by writing in the requested information.

1. I really like science and math courses.
   1___Strongly agree
   2___Agree
   3___Disagree
   4___Strongly disagree

2. I am not sure that college is worth all the bother.
   1___Strongly agree
   2___Agree
   3___Disagree
   4___Strongly disagree

3. My career plans after I finish college are:
   1___Very definite
   2___Fairly clear
   3___At the "maybe" stage
   4___Still pretty much undecided

4. What is your present year in college?
   1___First
   2___Second
   3___Third
   4___Fourth
   5___Fifth or more

5. Do you live at home, in a dorm, or where?
   1___At home with my parent(s)
   2___In my own apartment or house
   3___In a dorm
   4___In a sorority
   5___In a fraternity

6. During an average week, how many hours do you spend studying for college?_____(write in number)

7. What is your GPA (Grade Point Average)?_____(write in number) If this is your first term in college, report your high school GPA.

8. Do you belong to a fraternity or sorority, whether as a pledge or as an active member?
    1___Yes
    2___No

9. If you had to choose between a course in literature or a course in science, which would you probably select?
    1___Literature
    2___Science

10. What is your age?_____

11. Have you ever received a ticket, or been charged by the police, for a traffic violation—other than illegal parking?
    1___Yes
    2___No

12. Were you ever picked up, or charged, by the police for any other reason, whether or not you were guilty?
    1___Yes
    2___No

13. Have you ever shoplifted?
    1___Yes
    2___No

14. Whether or not you ever have drunk alcoholic beverages such as liquor, wine, or beer, do you do so now or are you a total abstainer?
    1___Drink now
    2___Abstain now

15. Have you EVER tried marijuana?
    1___Yes, in the past year
    2___Yes, but not in the past year
    3___Never

16. Have you EVER tried cocaine (crack, rock, freebase)?
    1___Yes, in the past year
    2___Yes, but not in the past year
    3___Never

17. During the past year, have you been nauseated or vomited due to your drinking or drug use?

1___Yes

2___No

18. Have you ever cheated on an exam?

1___Yes, very often

2___Yes, quite often

3___Yes, a few times

4___Yes, but only once

5___No, not ever

19. IF you have ever cheated on an exam, was that in high school or in college?

1___In college

2___In high school

3___Both

20. Do you agree that people ought to have the right to end their own lives anytime they are tired of living?

1___Strongly agree

2___Agree

3___Disagree

4___Strongly disagree

21. Do you favor or oppose the death penalty for persons convicted of murder?

1___Favor

2___Oppose

Do you approve or disapprove of abortions under each of the following circumstances:

22. If the woman is not married and does not want to marry the man.

1___Approve

2___Disapprove

23. If the woman's own health is seriously endangered by the pregnancy.

1___Approve

2___Disapprove

24. If the woman is married but doesn't want any more children.

1___Approve

2___Disapprove

25. If the woman simply wants an abortion for any reason at all.

1___Approve

2___Disapprove

26. Do you smoke?

   1___No

   2___Yes

27. Do you have a computer?

   1___Yes

   2___No

Do you think the government ought to spend more money, less money, or the current level on each of the following:

28. On welfare?

   1___Spend more

   2___Spend less

   3___Spend at the current level

29. On defense?

   1___Spend more

   2___Spend less

   3___Spend at the current level

30. On space exploration?

   1___Spend more

   2___Spend less

   3___Spend at the current level

31. On education?

   1___Spend more

   2___Spend less

   3___Spend at the current level

32. The press often reports predictions about the future by people such as Jean Dixon, who claim to have psychic powers. Do you think some people do have such powers?

   1___I am certain some people do have psychic powers.

   2___I think some people probably have psychic powers.

   3___I tend to doubt that anyone is a psychic.

   4___I am certain this is all nonsense.

33. How much confidence do you place in astrology—the theory that the position of the stars and planets in relation to our birthdays has a lot to do with what we are like and what will happen to us?

1___A lot of confidence
2___Some confidence
3___Not much confidence
4___No confidence

34. Whom did you favor in the 1996 presidential election?

1___Dole
2___Clinton
3___Perot
4___Other

35. About how often do you attend religious services?

1___More than once a week
2___About once a week
3___At least once a month
4___At least twice a year
5___Seldom
6___Never

36. What is your religious preference?

1___Catholic
2___Protestant
3___Jewish
4___Other
5___None

37. Would you say that you are a religious person or that you are not?

1___Very religious
2___Somewhat religious
3___Not very religious
4___Not religious

38. When you were in high school, did you participate in an organized sport that involved competition with other schools?

1___Yes
2___No

39. Are you employed?

1___Yes, full-time
2___Yes, part-time
3___No

40. What is your current marital status?
1___Single (never married)
2___Married
3___Divorced or separated
4___Widowed

41. When you were 16, with whom were you living?
1___Both parents
2___One parent and a stepparent
3___My mother
4___My father
5___Others

42. Thinking about your parents, or the people with whom you lived during high school, compared with other American families, would you say their income was below average or above?
1___Far below average
2___Below average
3___Average
4___Above average
5___Far above average

43. Thinking about the home you lived in when you were in high school, about how many hardcover books were in the house?
1___0–10
2___11–25
3___26–75
4___76–100
5___101–200
6___201–500
7___More than 500

44. What do you think is the ideal number of children for a family to have?
0___None
1___One
2___Two
3___Three
4___Four
5___Five
6___Six or more

45. Race/ethnicity:
   1___White (Anglo)
   2___African American
   3___Asian American
   4___Hispanic American
   5___Native American
   6___Pacific Islander
   7___Other_____(write in)

46. Sex:
   1___Female
   2___Male

47. Were you born in the United States?
   1___Yes, in this state
   2___Yes, in another state
   3___No

48. Do you agree or disagree that a preschool child is likely to suffer if his or her mother works?
   1___Agree
   2___Disagree

49. It might be better for everyone if the husband takes care of earning a living and the wife takes care of the home and the children.
   1___Agree
   2___Disagree

50. Do you usually wear one or more rings on your fingers?
   1___Usually wear more than one ring
   2___Usually wear one ring
   3___Sometimes wear a ring
   4___Seldom wear a ring
   5___Never wear a ring

51. At present, do you have your own car?
   1___Yes
   2___No

52. If you had to be one or the other, would you rather be a dog or a cat?
   1___Dog
   2___Cat

53. Would you rate yourself as overweight, about right, or underweight?

1___Quite overweight

2___Somewhat overweight

3___About right

4___Somewhat underweight

5___Quite underweight

54. How likely do you think it is that during your lifetime you will suffer a significant hair loss?

1___Very likely

2___Somewhat likely

3___Not very likely

4___Very unlikely

55. Have you ever used a sewing machine?

1___Often

2___Sometimes

3___Once or twice

4___Never

Please write in an answer to each of the following questions, estimating the answer to the best of your knowledge

56._____Estimate the world's total population (in billions).

57._____Estimate the percentage of the U.S. population who are African American.

58._____Estimate the percentage of Americans who are the victims of violent crime in any given year.

59._____Estimate the American family's median income in dollars per year.

60._____Estimate the percentage of Americans of voting age who voted in the 1996 presidential election.

61._____Estimate the percentage of the total popular vote received by Bill Clinton.

62._____Estimate the percentage of American adults who drink alcoholic beverages.